D0780446

DISCARD

A Gift From
ST. HELENA PUBLIC LIBRARY
FRIENDS&FOUNDATION

ALSO BY GRETEL EHRLICH

Facing the Wave

In the Empire of Ice

The Future of Ice

This Cold Heaven

John Muir: Nature's Visionary

A Blizzard Year

Questions of Heaven

A Match to the Heart

Arctic Heart

Drinking Dry Clouds

Islands, the Universe, Home

Heart Mountain

The Solace of Open Spaces

Unsolaced

Unsolaced

Along the Way to All That Is

GRETEL EHRLICH

PANTHEON BOOKS, NEW YORK

Grateful acknowledgment is made to the following for permission to reprint previously published materials:

Counterpoint Press: Excerpt from "Valley Wind" by Lu Yun from *Encompassing Nature* by Robert Torrence, copyright © 1999 by Robert Torrence. Reprinted by permission of Counterpoint Press.

Rifat Latifi: Excerpts from "Move above tree line . . ." and "Night has become my body guard . . ." by Rifat Latifi. Reprinted by permission of the author.

Library of Congress Cataloging-in-Publication Data
Name: Ehrlich, Gretel, author.
Title: Unsolaced : along the way to all that is / Gretel Ehrlich.
Description: First edition. New York : Pantheon Books, 2021.
Identifiers: LCCN 2020013919 (print). LCCN 2020013920 (ebook).
ISBN 9780307911797 (hardcover). ISBN 9780307911803 (ebook).
Subjects: LCSH: Ehrlich, Gretel—Homes and haunts. Ehrlich, Gretel—Travel. Authors, American—20th century—Biography.
Classification: LCC PS3555.H72 Z46 2021 (print) |
LCC PS3555.H72 (ebook) | DDC 818/.5409 [B]—dc23
LC record available at lccn.loc.gov/2020013919
LC ebook record available at lccn.loc.gov/2020013920

www.pantheonbooks.com

Jacket photograph by Joe Riis
Jacket design by Kelly Blair

Printed in the United States of America
First Edition
2 4 6 8 10 9 7 5 3 1

For Neal

Unsolaced

PROLOGUE

. . . I rhyme
To see myself, to set the darkness echoing.
—SEAMUS HEANEY, "Personal Helicon"

The ribbed hill, gray to green. Pronghorn grazing in first light. The folded mountains, two owls calling, five low-flying geese, and the near-frantic four-note morning call of the robin. I'm in an off-grid cabin set on a glacial moraine surrounded by kettle ponds where, ten thousand years ago, retreating glaciers left lumps of ice. Green rings every pond and the white folded mountains dive down to foothills threaded with blue flax, sagebrush, and native bunchgrass.

At dawn a calligraphic shadow—loose, wild, and precise, like old Chinese grass script—curtains the forested east hill. I walk inside it, then emerge in sun trying to re-create parts of my past, as when I once hid behind Wyoming sagebrush watching sheep graze when all this began. Would it be better to write nothing at all? No doubt it would. Yet here, I feel most at home.

Everything is moving, but there's so much we can't see: how thought comes into being; how grasses and trees connect; how animals know weather, experience pleasure and love;

how what's under the soil, the deep microbial empire, can hold twenty billion tons of carbon in its hands.

The mind splices fragments of sensation and language into story after story. The blood in my veins and every blade of grass is oxygen, sugar, photosynthesis, genetic expression, electrochemistry, and time. I watch clouds crush the last bit of pink sky. Breath slips even as I inhale, even as snow falls out of season and mud thaws, even as lightning ignites a late spring.

I try to calculate the time it takes to scratch these words. Thoughts flare and fade. Ink across paper registers a kind of time theft during which I fictionalize an ongoing present, the ever-elusive *me, you, here,* and *there,* all existing somehow in a slightly fraudulent *now*.

My cabin faces stacked peaks that reach 13,800 feet and are part of the Wind River Mountains. Those mountains are my mind's wall and wellspring. Down here, the light is peach colored, and as the sun shifts, one loose shadow, like thought, takes on a sharp edge.

Nearby, a meadowlark sings the western meadowlark anthem. Territory is presence. Presence means song, then nest. Nest means egg, fledgling. Time flies and stars are dying. I try to count the split-end strands of lost friendships spliced to new loves, betrayals and failures, houses built, lived in, and sold, as if nothing could possibly be held close or hold me motionless, as if there were no door I couldn't exit, no door that would let me in.

What has been forgotten, gone unnoticed? Stacked notebooks don't begin to frame it all, yet I page through them omnivorously, trying to catch a glimpse of myself and others, and the places we've lived in. How do we know anything? How do we lose it so easily? Almost daily I return to the high country. Mountain is shoulder: I rub against it and step forward. The hinge squeals, an arm lifts, a rock wall slides, and for a moment the mountain's inner sanctum is revealed.

Later, down-trail, I lie on grass in the sun with my horse grazing nearby and touch the frayed ends of memory, a soft mane of them, as if fingering braille.

*

In that eyelid of time between night and now, the horse whinnies at five thirty in the morning, startling a pair of sandhill cranes that have nested nearby. It's early May and the pond sucks green from the field, lays it on its surface like a coat. The sun's metallic sheen spreads between cattails whose million seeds have yet to burst. I get up and pace, sink back on the couch, walk up the hill, sit on bare ground between two muddy ruts. A whole day goes by. Night is no cushion. Nor is comfort. It's been snowing and raining here, and the mud deepens.

I've moved too much—something like twenty-eight times since I came of age—and I can't always anchor my spirit. But why would I want to? Anchor it to what? I've loved each place deeply. I try to imagine the comfort of sameness—those friends who live in the houses where they were born or to which they returned, and imagine too the discomfort it must arouse, the sense of confinement.

Sun comes out. Freedom is the green pond turning blue, the muskrat pushing dried reeds and grass to the far bank while making a summer house. I imagine hundreds of mud-and-straw huts clustered together, lit by glowing lanterns like the ones I saw on Kyoto's Kamo River. My own building project—a writing studio—is clamorously under way. Yet I'm saddened to see sawn and planed logs stacked up. If I listened, I might hear the chaos of those trees being dismembered, their bark peeled, their tendons sliced and the unbearable noise of nail guns assaulting their limbs.

Spring is this: One day the sandhill cranes dance; the next day swallows arrive and push bluebirds from their nest box. A week later an aspen leafs out, while on the mountain, beetles kill off every whitebark pine. Seventeen pronghorn antelope attempt a river crossing and are washed downstream, get out, try again. A prim gray cloud passes over. A porcupine sleeps in the willow, one leg dangling, black nose pointed up smelling for fire, smelling for rain.

Worldwide, violent storms split trees in half, persistent droughts suck bones, rain loosens whole mountains: a mud flow destroys my childhood home, a cornice crumbles, a typhoon drowns a hundred people in Japan, hurricanes raze Caribbean islands, a volcano blows, an avalanche takes three friends.

I rise early. Between first light and coffee, a pair of honking geese fly low over rising water. I hear the jake brakes of a semi loaded with last year's hay going downhill on the road. Rain begins.

Smarty, my horse (from the bloodline of Smart Little Lena), runs for his shed when lightning flashes—not a cloud-to-ground strike like the one that got me, but cloud to cloud—an infusion of savage energy. A goose calls, gets an answer, tips its head back in delight as its mate arrives, and the builder sashays through snowdrifts by dogsled. The generator is started: a table saw grinds through forests and a whole room goes up. In rain-light the mud glistens.

Uncertainty is on the rise. When hasn't it been? On the Big Island, where I live four months of the year, magma burbles up from Earth's mantle, lifts and falls like blood pressure. "Pele's lungs are big!" a Hawaiian friend says gleefully. "Look at her breathe!" The crater called Halema'uma'u collapses and fires steam-driven projectiles into the air. Fissures spew fountains of lava, red manes. And all down the mountain molten rivers are on the roll, on the run.

"Pele is cleaning house," Kawe says. "She is taking it all back." A photograph from the 1924 eruption might explain why: the local residents were sun-darkened Hawaiian farmers and fishermen. Now most of the evacuees are recent émigrés from cold states. "They wanted heat. They're getting it," a local at the gas station quipped. The mingling of cultures, of rich and poor, of urban and rural, goes on everywhere, like it or not. What matters is only how we live with others, how we learn from each other, how we thrive in our aloneness, how soon we bow down to Pele.

Becoming "native to a place" doesn't have to be about secured boundaries of blood and territory but can allude to a deep, growing knowledge of that place. The way one feasts on it and becomes nourished and gives thanks. And hands it over to be shared.

My DNA swab locates the sum of me in Austria, northern Scotland, the Jewish diaspora of Eastern Europe, Germany, and Finland. My paternal great-grandfather left the Austro-Hungarian Empire in the 1880s when Jews were pushed out of town; my maternal grandfather, the son of German immigrants, spent his early years with the Yankton Sioux and later became a lumberman in Minnesota. Pushed by war, love, famine, boredom, geopolitics, racism, and curiosity, we are always on the run. How do we remember what we were in so many different places? What betrayals and joys did we experience? What vegetables, fruits, soils, mountains, oceans, sunrises, and sunsets entered our senses and bodies, and what did they tell us? How did we learn to listen?

Our passages are marked not only by lash wounds, tattooed numbers, or stamped passports but also by genetic instruction from our DNA and the modifications to it made by the environment. We are shaped by environment as much as by our ancestors. The epigenetic memory of our cells is constantly being made. The push and pull, blast and bang of life is marking us in ways we haven't understood until recently. The grandchildren of survivors of the 1944–45 Dutch famine, brought on by war, have increased glucose intolerance. Some grandchildren of Holocaust survivors have altered levels of cortisol that makes them more easily traumatized than others. The transgene carried by nematodes affects their ability to glow. In a trial, when subjected to hot temperatures, they lit up. When taken to cool temperatures, their fluorescence stopped. But fourteen generations later their progeny, never subjected to heat, nevertheless glowed. The genetic memory persisted.

Memory over mind. Mind over memory, and the parsing of genetic mayhem, those are my tasks. Near my cabin the muskrat has moved on downstream, the porcupine has climbed from the tree, the cattails have turned green. Later the pond is neither green nor blue, but gray and perforated by raindrops that create seemingly endless concentric circles. Change is never nonchalant. The dynamism of soil, ice, climate, weather, and cellular structures has been underestimated. Peace is chaos singing.

I try to deconstruct memory—how proteins wash receptors, how dendrites change with each new thought. In the rush of cellular mutation, microRNA scratches intercellular messages into the bloodstream. It tells genes to take a break: it silences gene expression; it can cleave strands into two pieces and destabilize the messenger acids in our cells.

Late in the afternoon the muskrat returns and swims back and forth, pushing reeds. The mind swims laps, memory is cantilevered over genetic turmoil, and the writing goes on as if from unseen instruction, silencing, cleaving, and destabilizing words and thoughts, while the "hum" in me, the human, pushes fragments into the semblance of story.

Part I

1.

Forty years ago, I drove north on a two-lane highway that tipped up and over the lip of Wyoming at Tie Siding. A cowgirl galloped across open range to the post office, tied her horse to the hitch rail, retrieved her mail, and galloped away. I stopped the car to look: the grassland spread out wide, and three separate herds of pronghorn antelope ran in different directions. Sun poked through clouds, spotlighting bits of ground as if on a stage where an animal appeared, hit its mark, then wandered on.

As I drove, the embrace came slowly, ardently, and took me by surprise. Home has no walls, no ceiling, nor is its purpose to protect. Wind shunted the car from side to side; an ocean of wind and the smell of sage covered and carried me.

Then the day went dark, and I passed through badly lit, dingy two-bar towns, crossed alkali flats, stream meanders, and endless open range. All the way north were mountains, meadows, buttes, and river-carved bluffs. When the snow-bright Bighorn Mountains appeared in moonlight, I broke out laughing. It was May 15, 1975, and I knew I had come home.

Not that I had been looking for one. I'd never been to Wyoming, and "home" was not in my vocabulary. But the mountains and grasslands twitched and beckoned. There was no stopping. The road before and behind me seemed to merge under the tires as if to recover a memory not yet made.

Early on I saw how conventional society wanted me to be one thing only, reduced to a splinter in a reductive world, but I went the other way and kept unpeeling my mind. Sagebrush and string quartets, Buddhist practice and cowboying were all of a piece. Quantum decoherence interested me more than mapping out a firm life plan. The ground would always be spacious landscapes and animals. The sky would hold the soul-songs of Brahms and birds; the blue shawl of imagination would enfold everything else.

When David, my partner, lover, and co-filmmaker, died of cancer when we were both twenty-nine, all that we had before us, not only a new life together but also generous filmmaking and writing grants, disappeared. To stay in Wyoming wasn't a matter of hiding. I was living my grief to the hilt, soaked in it while working outside with animals and with people to whom I didn't need to explain. The book I began writing then wasn't about personal loss but about what I had found: what and who saw me through, and who saw through me.

One doesn't "get over" a death: it stays with you forever, and at the same time, it sharpens desire. Like a wolf eyeing a herd of antelope, I made passionate forays into the cowboying life with a nothing-to-lose attitude. My friends in New York and LA asked when I was going to stop hiding. But I was living, best I could, and didn't want to go back. One survives or not. It didn't seem there was a choice.

Where does solace reside when you have lost it all? I was desperately poor and had no real winter clothes, not even a parka. But my dog, Rusty, and I took our daily walks under the great cottonwoods along the frozen North Fork of the Shoshone. The day we stumbled across a warm stream trickling in, I knelt down and picked watercress and miner's lettuce, fed some to him, and together we ate our fill of fresh greens. Once the snow set in—three feet on the level—there was very little going in or out. The lane into the cabin had long since drifted in, so I trudged up to the road and hitched rides into Cody's grocery store, then another ride back out, though by then all the fresh vegetables in my bag had frozen.

After New Year's Eve the temperatures dropped. Nights were minus fifty and daytime temperatures never rose above twenty below. I helped a rancher feed cattle stuck in the mountains when it was minus fifty. Come spring, I bought a sheepherder horse from the local horse trader for a hundred bucks—an old Appaloosa named Blue sired by the last government remount stallion from the days when the federal government helped farmers and ranchers breed their mares.

When the driveway was finally plowed, I heard about a place for rent with horse pasture in a small town called Shell across the Bighorn Basin. A friend's aunt "Mike" and uncle Frank lived there. He told them I was coming, and loaded Blue, Rusty, me, and my belongings into his truck to make the trip to my new home. Walls of melting snow lined both sides of the road. We arrived on Trapper Creek two hours later under the first blue skies I'd seen for months.

It was an old log house with an enormous cottonwood out the kitchen window. My nearest neighbors—none were visible from the house—were the local equine veterinarian and John, a modern remittance man from the East Coast who was tending his family's small ranch. The soil was red, and I wanted to put in a garden. When I rode Blue to town to get my mail I stopped where John was planting. "Want a partner in that?" I asked, pointing to some unplanted rows. He said, "Sure." And the next day I walked to the field, seeds in my pocket and shovel over my shoulder. Later, a handsome young rancher rode by looking for strays. He wrote his name and phone number in the dust of the road and told me to call him if I saw any.

That night the phone rang. It was my friend's aunt Mike: "I hear you want to cowboy. We're moving Stan and Mary Flitner's cattle to the mountain. I'll pick you up at five a.m. Don't need to bring that sheepherder horse of yours. I'll bring you something decent to ride."

Couldn't sleep. I fastened my spurs to my boots, laid out fresh clothes, and retied the yellow slicker on the back of my saddle, then walked up the lane to look at the stars. The winter had been so long and hard that the sky in spring was like a reservoir: not fresh water, but fresh light.

To the east, the Bighorns rose on stilts of red walls that layered up to thick alpine forests. The Big Dipper was engraved on the black page of the sky, its cup brimming, as if to say, "We have plenty here." Rusty sat in the middle of the road, tipped his head back, and howled.

Loneliness is a lie of ego, though sometimes the absences mount up into an ache that can't be ignored. Loneliness ended that night, whooshed away by the embrace of strangers. Unknowingly I had entered an obscure ranching community in which I could keep a writer's distance but still be included. I could "cowboy" and write. Shell became my heart's home.

*

Every morning Mike arrived wearing fresh clothes: a starched shirt, silver earrings, clean Levi's, boots, spurs, and a Silverbelly Stetson. She asked me to drive and haul our horses up the mountain. That became my payment for her mentorship. On the way up she gave me advice: "I know you can ride, but just remember, these aren't sheep. Stay on the downhill side of the cattle and push them up off the creek. I'll ride along with you."

I don't remember much of that first day except that Mike was always by my side quietly instructing me on where to be and when and why. "Don't wait for the worst to happen. You have to anticipate what the animals are going to do," she told me. "You have to watch them carefully. When going up a steep mountain, the calves will stray behind the mother cows and try to run back; sometimes they succeed, and we have to bring the mother cows all the way back down and start over."

At the end of the day we hauled the horses to the bottom of the mountain and stopped for a drink and a hamburger. Mary said she thought I was a hitchhiker passing through, and in the larger sense I was, compared to the third- and fourth-generation ranchers around me. I told her I was a newcomer, had moved to a house on Trapper Creek, and was working on a book. "What's it called?" she asked. *"The Solace of Open Spaces,"* I said. Mike smiled. "Well, you'll have to come back up the mountain tomorrow, then. You'll need fresh material."

In that quiet, understated way, the welcome was heartfelt, but I knew I'd have to earn my spurs. Explanations were not necessary—no need to talk about the past. We moved cattle slow and steady, gently but firmly, though of course there were horse wrecks, buck-offs, tangled ropes, all kinds of small disasters about which we laughed at the end of the day.

As the newcomer, I was the comic relief. Once I jumped into the fast-moving creek to rescue a calf, but by the time I caught up, the calf had climbed out of the water and was back with its mother. Soaking wet—boots, chaps, jeans—I climbed up the bank as the other riders sat on their horses, laughing.

To be teased meant I was accepted. It was the lubricant of camaraderie; spring roundup meant that winter blues were banished. Just the way I wanted life to be. Not denying the grief that hung on but simply giving it space, an opening into the next day.

. . .

Mornings at dawn, I rode the green swale that folded up into forested mountains with pocket meadows all the way to the top. I didn't have to talk. Solace hummed, my saddle creaked, spurs jingled, chaps were zipped all the way down my legs. I bought a good horse. My dog was at my side. We worked horseback long days—sometimes ten, twelve, or fourteen hours in sleet, hot sun, hail, rain, wind, and snow. No longer strangers, we looked out for one another, anticipated the needs of the moment. Back straight, hat low, rein-holding hand over the saddle horn, I acquired an on-the-spot view, not of what I had lost, but of what I had gained.

Solace was there and some days, everywhere. Other days, it was nowhere to be seen. That's when I'd find it cowering inside a long stretch of solitude, or along a dusty trail with cows ahead of me and no water, like the day Stan tied a banana onto my saddle, tight enough to give the banana a "waist," and said jovially, "Lunch. Made it myself."

Stan loved practical jokes. He'd ride by, undo the throat latch of the bridle when I wasn't noticing, then take the bridle off and slap the horse on the rump: off we'd go, but never far. Just far enough to provoke laughter. In early winter, when we bundled up in insulated coveralls, he'd offer to give me "a leg up," then throw me all the way over the horse. It didn't hurt. I was young and the clothes were well padded.

Sometimes unexpected laughter was all I needed. Extremes of weather and horse problems generated humor. On those days a tiny hint of solace shone through. It could be nothing more than the odd-shaped shadow cast by a pebble on the ground as a last ray of sun struck down despair.

There had never been a question of leaving. I wanted to stay exactly where I was. I rode hard, danced hard, and eventually "mothered up" with a few young cowboys and took them to bed. Grief pricked and tore at me; strenuous living balanced out the emotional turmoil, and I hoped that a life with animals and physical exhaustion would hold me steady. It would have to, because even when I tried to leave—when I packed up and started south—I'd come to a sudden halt, pull off the road, and before I knew what I was doing, turn around. Sometimes you have to leave; other times you can't leave no matter how hard you try.

2.

From the first day in Shell, Mike had been my mentor, and more than that: she and her family became my family, an all-around love affair. High-strung, poised, and sardonic, Mary Francis "Mike" Tisdale Hinckley had nervous hands and soft blue eyes that narrowed to slits when responding to some rude remark, but her cheeks, like little suns, were always rosy. She had a delicate nose, wore lipstick, and looked elegant on a horse no matter what was happening. She never lost her cool. Long-legged with a gray braid or ponytail down her back, she was always in the right place at the right time. I learned from watching her when and where to turn the herd.

Much to everyone's surprise, she didn't marry a big rancher but rather a generous, smart, funny, storytelling highway contractor, Frank Hinckley, whom she'd met at a dance in Kaycee. "Those other sonsofbitches just wanted me to stay in the kitchen with the kids," she told me. "But Frank, he was different. He was all for my cowboying life. He respected me."

Evenings, she sat on a high stool at the kitchen counter drinking her nightly cocktail while Frank pan-fried steaks. He was devoted to her and in awe of her cowboying prowess. Her string of well-bred quarter horses could be a handful, but she was always self-assured, keeping calm when things got "western." Nothing escaped her; no one could dodge her keen wit, especially the men and their foibles. Yet her courtesy was part of the cowboy etiquette she had learned on the remote ranches in the Hole-in-the-Wall country where she grew up and cowboyed as a young woman, and her sense of right and wrong was honed by her belief in equality for all. Gender never factored into her thinking. She was made the foreman of 125 sections of land that made up the TTT Ranch when she was in her early twenties. Yet she expected to be treated like a lady.

When Mike left the University of Wyoming to work on the ranch, her father admonished her: "You better make a damned good hand of yourself," and she did. That first year she didn't know how to rope, so her brother Tom decided to teach her a lesson: "We gave her the ground job. We were delousing cattle, and by the end of the day, she asked for my rope. She practiced out of sight for months. By the time we had to work cattle, she had taught herself to head and heel as well as anyone." Ever after she said, "If it can't be done horseback, it ain't worth doing." She groaned if anyone went into the branding pen on foot. "Damned foot soldiers," she'd mumble under her breath. But a great hand she became and was posthumously inaugurated into the Wyoming Cowboy Hall of Fame.

"The red-wall country was terribly isolated," she told me. "We called it Siberia. Summers it was hot, full of rattlesnakes and quicksand on the Powder River. Winters, when the cold settled into the valleys, no one went anywhere for two or three months. But when it lifted, boy, did we go to town! There were dances at the Occidental Hotel in Buffalo and the Grange in Kaycee."

In the winter of 1949 when the temperature dropped to fifty-one below zero, she said they lost very little livestock. "We'd had a dry summer, so we trailed fifteen hundred yearlings up to Bucknam in five days. That made a full trainload to Omaha. I went with the cattle. Women didn't do that in those days. When I got back some of those green-eyed women in town asked which of the cowboys I'd slept with and I said, 'All of them, but I liked the horse wrangler best.' "

Mike's family had come north from Elgin, Texas, with the trail herds in the 1880s—part of the migration so wonderfully described in *The Log of a Cowboy*, on which Larry McMurtry's *Lonesome Dove* was based. That cowboy legacy included honesty, straightforwardness, and humor, plus a kind of Texas chivalry that went way beyond a man opening a door for a woman. Cowboy ethics consisted of hard work, modesty, humor, and a deep etiquette that informed the wildest out-on-the-range behavior. Mike's was a horse tradition, in which, she said, "a man would get on his horse to ride a hundred yards just for a glass of water," not out of laziness, but because the horse and the human were locked together inseparably.

Mike never raised her voice around an animal. She kept her steely toughness under her Stetson and used it only as needed. Modesty and a quiet way with livestock were part of the inherited legacy. It was as much a part of the old-time cowboy way as it is part of today's "natural horsemanship."

One November, before the snows came, she bought me a rope at King's Saddlery in Sheridan. "Practice every day. You'll be first up at the branding in April." I roped chairs and logs—anything I could find. I made loops, took my dallies, saw how quickly I could recoil the rope and shake out a loop, and finally started pulling things over.

At the first branding, I rode into the corral with all the neighbors watching. Terrified I'd disappoint Mike, I made a loop, threw, and caught a calf. No one was more surprised than me. "Take your dallies!" her son, Mart, yelled. I made a quick wrap and dragged the calf to the fire, then caught another and another. Mike was at the rail watching. She didn't say anything, just gave me a smile.

Later that spring Mike and I rode colts together, trotting out of the yard two by two, close enough that our legs touched so the colts didn't have time or reason to be afraid and buck or run. That year Mart won a tag to hunt mountain sheep in the mountains above Meeteetse, and he asked me to help him pack in.

We crossed the Wood River, then headed up into the mountains and made camp in an aspen grove. Somewhere along the road I lost my sleeping bag, so we made a bed of horse blankets and used Mart's bag for a cover. One thing led to another and I stayed. It wasn't the hunt that was important to Mart, but the joy of camping in the mountains with horses, and we had a fine time going for long rides in Indian summer weather. When we finished off the bottle of Irish whiskey, Mart declared the hunt was over, and we arrived back in Shell hand in hand.

Mike came out when she heard the trailer. She looked at the two of us and couldn't help smiling. Then she informed us that we had a cowboying job on the Crow reservation near Lodge Grass, so we'd better wash our clothes, change horses, and load up. We left with six horses at dawn and drove north into Montana.

The ranch was leased from the Crow Indian tribe. Headquarters was a modular home. The ranch had no roads, no cross fences. Our job, along with nine other riders, was to gather seven thousand head of steers. We had a month to find them all.

Morning call was at three a.m. Our mounts for the day were roped out of the remuda using a one-swing Houlihan. We had to be saddled up and ready to ride by four. The foreman rode out at a hard trot and we filed behind him in the first dim light, sometimes riding for two hours before we found any cattle. The leased land was thousands of acres of rolling hills with wild plum bushes in the coulees. Every day we gathered what we could and turned the animals for home.

Yearling cattle are like teenagers—difficult to keep track of—and once we had them strung out, there were cattle as far as the eye could see. Some days I was given a bunch to bring back, though I wasn't always sure how to get back to headquarters. I loped back and forth, turning the lead steer, then gathered the ones that had strayed behind.

Dinner was served at four in the afternoon, then bed shortly after, and we were up again at three a.m. One of those mornings in the arena where we saddled up, I put my foot in the stirrup and the horse took off bucking. I was able to swing the other leg over and ride him before I was bucked off. Miraculously I landed on my feet. There was applause. I laughed at my good luck, pulled my hat down, and straightened up while trying to figure out where I was. Mike led the horse to me, held him while I got back on, and together we rode out of the arena. Every day we gathered what we could and trailed "the gather" back to headquarters, where there were brand inspectors and semitrucks waiting. We changed horses to work livestock in the sorting corrals.

When all seven thousand steers had been accounted for, Mart, Mike, and I hauled our six horses back to Shell. There was fall work to do and we helped the Flitners bring their cattle down the highway out of the Bighorn Mountains as snow fell.

Winters in the 1980s were almost as ferocious as those a hundred years earlier, when so many cattle perished that, according to Mike's uncle, "Hard Winter" Davis, you could walk on cowhides from Kaycee all the way to Casper.

As soon as the snow had been on the ground for a while, Mike would get "cabin feverish," call me, and suggest we "go on a tour," which usually meant driving over the mountains to Sheridan to see her old friends. Blizzards and icy highways didn't stop us. I was expected to drive and did so, often terrified, but we made it and, once in town, went straight to the Elks Club, where we could get a good stiff drink and visit with her old friends—hired hands and ranchers from Sheridan, Buffalo, and Kaycee.

Those years after David's death weren't easy. I was broke and sometimes lost, wondering why I had chosen to live in a tiny Wyoming cow town, or if I had chosen at all. Yet Shell embraced me, and a community came into being: ranchers and outsiders, lovers and friends—a motley eccentric crew. We had midwinter dances at the Shell hall, and brandings and birthdays, weddings and funerals. Word of mouth worked: there were no cell phones, no internet, and no mechanized ranching. Inspired by Mart and Mike, my ideal was to live on a "cowboy's ranch," one that had no roads, where all the work was done horseback.

When Mart's and my affair ended, I called Mike and said I hoped she wouldn't be upset. She didn't comment, but sensing I was distraught, said, "Better come on up for dinner." Of course, Mart was there too. That was how the family operated: broad-shouldered, all-accepting of human foibles, refusing to sweat the small stuff. Friendship came first; western hospitality went all the way to the bone.

Mike and her extended family had been shaped by what the historian John W. Davis called the "shocking and murderous events of 1891 and 1892, which have come down to us as the Johnson County Cattle War." On December 1, 1891, John A. Tisdale, Mike's grandfather, was shot in the back while driving his buckboard home full of Christmas presents for his family. Frank Canton, hired by the local cattle barons who were trying to shut down the small-time settlers, had been lying in wait for him. After killing Tisdale, Canton shot the horses and the puppy Tisdale was bringing home as a surprise.

The Johnson County War was a class conflict between elite cattlemen and the less well-off who had filed on homesteads. The area encompassed the entire east flank of the Bighorn Mountains and the grasslands that spread out from it. Initially set aside as unceded Indian territory used by the Lakota, Cheyenne, and Arapaho, the rangeland was "opened" as soon as the Native Americans had been pushed off for white people to grab. And grab they did, not only from the Indians, but from poor cowboys. The systemic thuggery that benefitted the elite forever changed what had been an open and democratic society, with unfenced land free for anyone to use in a harsh but breathtakingly beautiful part of the country.

Mike and her brother Tom were seared by what the locals call "the Invasion." The summer we stood in the gulch where their grandfather was shot, tears flowed. Mike's usually calm demeanor cracked. I saw her deep indignation, and how a longing for justice permeated her life, how it shaped her psyche, how her resilience and toughness came from that deep wound.

The democratic open range of the West that Jefferson had earnestly tried to establish and de Tocqueville had written about, where there was room enough for everyone and grass to spare, had come to an end. Murder and corruption continued to be part of the American way.

"Not much has changed, has it?" Mart said not long ago when he, his brother, Mikey, his sister, Timmy, and I visited the ranch where Mike had lived. "The rich who are in power in this country still believe they are above the law," Mart said. At the Invasion Bar in Kaycee, the talk is still about the miscarriage of justice that allowed the killers of John Tisdale to go free.

*

When Mike first had heart problems and her doctor told her she needed to walk, she was horrified. "Walk? Me? On Horse Creek? They'll think I lost my horse somewhere!" But walk she did and hoped no one would see her. She was stubborn about maintaining the old western ways, not because they looked good but because they were the right ways of doing things, and she was firm in the understanding that whatever you didn't know, the land, weather, cattle, working dogs, and horses would teach you.

Mike's health began to decline in the mid-1980s: first breast cancer, then a heart attack. She recovered and thrived, but a few years later there was a second heart attack, and when the call came from the local hospital, I chained up the pickup, navigated through terrible mud down the steep ranch road, switched trucks, and hurried to town. By the time I arrived she was sleepy. I sat on the edge of her bed. One eye opened: "Can't you sneak me a cigarette?" she asked. I shook my head. "I'd bring your horse into the room, cook you a steak, mix you a drink—anything but a cigarette." She smiled, nodded, and was finally able to sit up. I brushed her long silver hair and braided it as she talked about things we'd rarely if ever discussed. She asked how my marriage was faring. My shrug said it all.

Mike and Frank's absence from the impromptu wedding had been conspicuous. Now I knew why. She said only, "Women don't want to be treated like men. We should always be treated with respect. We just don't want to be left out and there's no reason we should be. Riding with the men doesn't mean you have to become one."

To the end Mary Francis Tisdale Hinckley was wholly feminine. By that I mean: strong, vibrant, valiant, loyal, tenderhearted, and beautiful. And so modest, one might not have known what an accomplished horsewoman she was.

After an hour-long visit, her eyes closed. She said she wanted to sleep. I promised to be back in the morning. Maybe Mart and I could bring her favorite horse to the window. As I started for the door, she whispered, "Don't let the bastards get you down," meaning men, then let out her old cowboy yell: "Powder River. Let 'er rip."

She died the next morning before I could get to town. Her funeral was at the Shell hall. Five hundred people came from all over Wyoming and Montana. I gave one eulogy and Stan Flitner gave the other. He said: "Mike did everything with style. She was a classic horsewoman, a wonderful combination of rawhide and lace."

3.

The first time I saw Hudson Falls Ranch in the mountains above Shell was from horseback. Mart and I had ridden the edge of the mountains looking for strays. We had scrambled up over a rocky lip and came to the bottom of a high-altitude hayfield. Beyond was a small lake, another hayfield, and an old two-story house tucked into a cirque of mountains. I'd entered a secret hanging valley, a Song Dynasty landscape painting, and sat on my horse, staring. "Who owns this?" Mart said. "I think it may be for sale." It was getting late and we had to ride two hours back to his place. I tucked away that view of a mountain ranch and held on to it during the months to come.

In the winter of 1980, I met someone who could handle a team of horses, pull a pack string, and talk about nineteenth-century landscape painters and William Faulkner's novels. We were compatible and quickly partnered up. My few intense months with David years earlier and all that they portended were gone forever. I knew that living with anyone else would not be the same, but it seemed much too self-consciously romantic to retreat. "Professional widowhood" wasn't for me. I wanted to get on with life.

Our impromptu December wedding took place amid snowdrifts with a few friends, and the next winter, with the small inheritance my grandparents left me—the ones whom I'd told of my wish to become a writer and a rancher—we bought the end-of-the-road mountain ranch I'd first seen with Mart above Shell.

Hudson Falls Ranch was an old-fashioned, hardscrabble outfit—hard to get to, and I liked it that way. Winters, we fed hay with a team of horses and wagon. By four in the afternoon, as it got dark, the temperature dropped to twenty below zero. Chores had to be fast and exact. I checked cattle on horseback or skis, put "calvy cows" into the corrals near the calving shed, fed horses, fed dogs, split wood for all the stoves. We burned ten cords of wood every winter.

The first spring we bought fifteen bred heifers from friends. The first calf was "backwards"—a breech presentation with the tail and back hooves showing first. I panicked, ran to the house, called Stan in the middle of the night, and asked him to quickly talk me through the procedure. "Pull it steady but fast enough so it doesn't drown in there, then hang it to a rafter by the back legs so the lungs drain out, and after, hold it to its mother until it sucks."

Back at the calving shed, we wrapped the calf-puller's chain around the two back legs and steadily winched it out, not too fast, not too slow, going with the breathing of the cow. The calf was alive. I rubbed its back, cleared its mouth with my fingers, and watched as it found its mother's milk. After that initiation, the rest of calving went smoothly.

It was winter country up in the mountains and I was the night calver. I checked "calvy" cows, stuffing my flannel nightgown into insulated coveralls, threw on a parka, then skied to the cattle every three hours from ten p.m. until dawn.

Ardor and *arduous:* those were the two words that defined calving time. The work was sometimes difficult, but the nights blazed with occasional northern lights and a thick mat of stars. I could ski quietly into the herd and not disturb them. I could hear the night sounds of coyotes and owls. I was sleep-starved, vigilant, and watched every calf being born. Each one arrived with a distinct and unique personality and an inborn natural beauty. Still wet, they shook their heads and looked fervently for the tit. In my calf book I marked down the length of time it took for the calf to get up and suck and how well the mother cow helped out. Negligent cows were sold in the fall. That's how I built a self-sufficient high-altitude herd of cattle.

Hudson Falls Ranch had been homesteaded in the late 1800s. The ten-by-twelve-foot cabin where the homesteaders had lived and given birth to their first child was still there, but everything, old and new, was run-down. No corral gate swung, no fence was in order, and the hundred-year-old house was uninsulated. We found wooden barrels full of old nails that we straightened at night on an anvil. From a photograph of the ranch in homestead days we began the arduous work of re-creating it as it had been, lifting cedar rails and posts out of the mud and remaking sorting corrals. We tightened barbed-wire fences, cleaned irrigation ditches, fixed up a shed for a tack room and another for firewood, then cleaned out part of an equipment shed to make a calving barn.

Every room was heated with wood. I cooked on a wood cookstove that had belonged to Buffalo Bill, baked bread and pies, made soups and stews, though I was less interested in the history such a stove represented than in keeping the house warm.

The horse and cow herd gradually increased. We kept every heifer calf—females that would become mother cows—and built the herd slowly, the old-fashioned way. Mike gave me a bred heifer who was part longhorn. I called her Spot and she'd come if I called her. We bought a buckskin and a bay horse from the Wind River Reservation, and later I bid on a long-legged sorrel gelding with Doc Bar and Doc's Jack Frost bloodlines out of a Texas-bred broodmare. I bought a female kelpie dog to breed with Rusty and kept two of the pups—Sam and Foxy—who were brilliant at their ranch job of moving cattle.

The hundred-year-old house was made of gypsum block. It was like living inside an aspirin. The block crumbled and the cold poured in. I had to warm up the bottom of the bathtub with a space heater to keep the water from freezing when I turned on the valves. Days I wrote. Nights I read huddled by the woodstove in the living room with gloves on. There was no television and no NPR. I could see my breath, and the aging pipes, wrapped in heat tape, often froze. To my parents' eyes (they visited once and that was enough for them), it appeared that I had drifted into poverty. My mother said, "Why didn't you buy a nice ranch in Jackson Hole?"

They'd already admonished me about the confinement I'd experience with so many animals to care for, but I paid no attention. When I finished writing *The Solace of Open Spaces* and the first money I'd ever made from writing came my way, I bought a semi load of cattle, a new refrigerator, a Ford pickup truck with a straight-six engine, a team of Belgian horses, a pregnant donkey, and another kelpie, who soon had a litter of four puppies.

For months at a time we rarely left the ranch. Winters were tough, cold, and snowy. I wore a red L.L.Bean union suit, Wranglers, a wool vest, a canvas jacket, a silk neck scarf, a Scotch cap, and Handy Andy gloves, plus five-buckle over-boots over my cowboy boots. Nothing we wore kept us warm. But since there was nothing better, we laughed at our misery and watched out for hypothermia.

Solace comes in unexpected ways. The limitations of ranch life, what Mike had called the dawn-to-dark "can't see to can't see" predicaments of daily life, became a kind of liberation. What I had asked for as a child—a ranch and a life of writing—I now had. The mountains rose up straight out of the back pasture, its cirque folded around me. With a good library and bountiful animals, I found my aversion to "home" making a sharp turn. Staying in one place and going deep broke through my restlessness and enlarged all that I saw. I looked around, and what was around looked at me.

*

On the midwinter night that my dog Rusty refused to come into the house at dark, I began to understand what an old cowboy I'd met at a bar in Cody called "ranch ESP." Rusty was sitting atop a manure pile looking south toward the lower field and wouldn't come in. That was unusual for a cowdog—he was always at my heel. When I went to get him, he whined and jumped up, nipped my arm, and led me through the lower pasture for a quarter mile. At the fence line I saw a coyote chewing the back legs of a newborn calf that had frozen to the ground. Rusty chased the coyote away. My husband soon arrived and, seeing the problem, took off his boots and wrapped his socks around the calf's back legs. We carried the injured animal all the way back to the calving barn with the mother following. The calf lived.

Rusty had shown me the way. "The way" was right there under my nose, ludicrously close. I only had to pay attention. On another night in the middle of dinner, Rusty looked nervous. A snowstorm turned into a blizzard, and I knew something was wrong and found myself walking through deep snow to the horse corral. Snow had drifted into the three-sided shed all the way up to the horses' shoulders and they couldn't move.

Digging them out took hours. A little longer and they would have suffocated.

One autumn night under a full moon, Rusty clawed my arm and led me to the hill above the house, eager to show me something. A herd of elk had bedded down in the hayfield. We lay on our bellies at the lip of the hill and watched. The wind was right, so they didn't smell us. Rusty looked at me, looked at them and back at me, so pleased that they were there, part of the family.

The whole-body sensorium is alive in each of us—dog body, horse body, elk body, human. Sentience is as much about perceiving what others know and are, as it is about examining our own psyches. Knowing "otherness" feeds into who we are, and the shared communication binds us: me to Rusty; Rusty to the elk; the elk back to all of us. Sniffing, stamping, running, sleeping, watching, seeing—all converged. Unconditional living—being alive was pure pleasure.

Another wet spring when I was alone on the ranch during calving, sickness broke out in the new calves: quick pneumonia and scours. I worked day and night giving shots, dragging calves out of puddles of water. One morning at five, a car pulled up. The ranch was at the end of the road and I couldn't imagine who it could be. I heard whistling, then Erv appeared.

"Heard you were having troubles up here. Let's get to work." How could he have known? I'd told no one, so I chalked it up to ranch ESP. He split wood for the stove, refurbished the vet kit, and out the door we went to save lives. We moved the small herd to fresh ground, fed them straw and hay, gave shots to the sick, and saved all but one calf. At the end of calving he refused money. "I didn't come for the money. I came to help," he said, and went home.

. . .

As quickly as the snow came, it melted under a hot sun and the meltwater roared down from the mountains. I set the headgates and laid out forty-five tarp dams. To get water over as many acres of hay ground as possible was important, but that meant I had to change dams three times a day.

In the large fenced vegetable garden, I planted rows of every kind of green bean, plus spinach, chard, beets, carrots, squash, lettuce, radishes, cucumbers, peas, and broccoli. What I didn't consume in the summer months, I blanched and flash-froze for the winter.

There were four large chest freezers—two for meat and fish, two for vegetables—and we kept them full. Trips to town happened infrequently. We kept two 4×4 pickups with chains on all four tires. The truck at the bottom was for the highway.

From late June to October my husband led pack trips in another part of the state, and I managed the ranch alone. Planned grazing meant moving electric fence, salt, and cattle every three days, irrigating hayfields, whelping puppies, helping neighbors, planting and harvesting a large vegetable garden, fixing fence, and writing. Too busy to be lonely, I would have welcomed help, but there was none.

Part of surviving on my own was to learn what we called "the widow-woman's way." From Bud Williams I learned to work cattle quietly, without stress. I threw the hotshots away and rebuilt the sorting corrals to eliminate the animals' fear, making dry runs with alfalfa cubes and hay on the other side of the squeeze chute as a reward, so that when I had to bring a sick cow in to give her a shot, I could do it on my own.

Ranching alone is neither fun nor romantic. I often lost my sense of humor and was scared. A bull I'd bought, which unbeknownst to me had been tranquilized, turned mean and chased me across the sorting corral. I ran for the fence, but sensing I wouldn't make it in time, I turned around to face him as he charged and yelled a deep piercing yell. The bull stopped, surprised. I ran and jumped onto the fence just in time to avoid being trampled.

Animals are big and I'm small. My own cows knew me and were gentle, but other cattle could be unpredictable. One neighbor's cow we'd taken in attacked me, knocked me backward, and butted my head. I tried but was unable to edge out from under the bottom rail. My husband rescued me, but he left for a family meeting in Georgia the next day. The bones around my eyes were broken and I had two cracked ribs that made it painful to split wood and load the pickup with hay to feed the cattle, so I called Mart for help. All summer his mother, Mike, insisted I leave a note on the kitchen table whenever I rode out to move cattle, indicating which pasture I'd ridden to, and when I got home, I was to call her. That's the kind of friend she was. It made being alone on an isolated mountain ranch for four months of the year bearable.

As soon as the fall work was done—collecting electric fence and tarp dams; oiling headgates; winterizing pickups; stocking the larder with months' worth of honey, flour, sugar, and dry food, as well as the last harvest; blanching and freezing vegetables—I had more time to write.

My writing room was a chicken coop with a cheap woodstove and no electricity. I wrote a novel on a small blue Olivetti typewriter and cut and pasted it on the floor, then retyped it. Every day in the winter the dogs would come to my writing cabin at exactly the same time: three thirty. They knocked over my skis, and off we went through the pastures, down the frozen irrigation ditch, or across the iced-over pond. By the time we returned at four thirty, the sky was dark, and I checked the livestock and made sure the water gaps were open.

The inward life and the outward became enmeshed. Ideas flashed through the flakes of hay I pushed off the back of the wagon. My hands still know how to tie the baling twine with a loop at the top to hang on the wagon's stanchions while simultaneously keeping track of thoughts that had sparked while my fingers did their job. I kept tabs on coyotes and elk. Keyboard and hay wagon, tarp dam and flow of mind began to seem the same. My friend Malcolm Margolin once said, "Our responses to Earth are dug in deep and are old. We aren't new to what we do and know. The old ways are embedded in us."

4.

I wasn't born to ranching, but I was born to horses. Every New Year's Eve, in the middle of my parents' annual party, my father would burst through the double front doors leading the stallion into the living room. My mother feigned anger: "Oh, darling, take that horse out of here," she'd say, and with a theatrical gesture, turn back to the guests, her eyes twinkling, and refill her glass with champagne.

Horses surrounded the house: broodmares, kids' horses, a palomino quarter horse, and several American saddlebreds that my father exercised in harness behind a sulky. He had always wanted a ranch and looked for one in Mexico where he had a thriving business in what was then called *la ciudad*—Mexico City—but he never settled on a place there or anywhere else. Perhaps he sensed how confining it would be, since he and my mother, who adored each other, also liked skiing, sailing, flying, traveling, and giving parties.

As a newlywed in Harvard, Massachusetts, he bought a horse and a sleigh as a surprise, and he and my mother went visiting in and around the village the old-fashioned way: "Horses with bells on," my mother would say. She had high cheekbones, bright blue eyes, and a presence; she dabbled in the fashion world, designing clothes and modeling, and was always elegantly dressed.

My father graduated from Cornell as a chemical engineer (with a minor in animal husbandry) and, when the war came, was sent to California to build a factory and make plastic tubing. He was the first to make the flexible tubing that is now so common in hospitals; he also made plastic fuel lines for the warplanes headed to the Pacific, replacing those made of rubber because plastic was lighter and less flammable. It was said his inventions saved thousands of lives.

When my parents had enough money, they designed the first of three houses. It was simple, modern, and accommodating, with sea-grass rugs and in-floor radiant heating, which my father designed and installed. One side of the house was all glass. Every room had a fireplace. The house curved around and opened out onto a patio, gardens, a wide freestyle lawn, and an ancient oak tree. There were corrals, pastures, a three-stall barn, and an oblong track for exercising horses. As a teenager, my father had been sent to Wyoming to work on the Triangle X Ranch. He cowboyed there and rode into Jackson Hole, threw dice in a back room at the Wort Hotel, helped pack the buckboard with supplies, and rode back to the ranch in Moran.

Blue-eyed and black-haired, he was a charming bon vivant with an inventive mind who saw no obstacles, only problems that could be solved. He'd watched Charles Lindbergh do touch-and-gos at an airport in his native Chicago and decided that flying a plane was for him. He always had one and spent lunchtimes getting in his airborne hours. My mother, sharp-minded and always a good sport, copiloted. Both parents had a sense of fun and a sense of humor. My sister and I were teased and made to do chores, and the night we "borrowed" a sailboat from the harbor, our father came looking for us in the Coast Guard cutter. After, he made us do the brightwork on his sailboat until it shone.

As parents they could be hypercritical. They locked me out of the house if I came home late from a date, which no doubt spurred my obsession with open spaces and my sometimes obstreperous "Don't Fence Me In" attitude.

I suspect my father's one unspoken regret was never having bought a ranch. That life would have been more than perfect for me. To be surrounded by horses and animals was my father's deep and unrealized longing, and though I didn't know it at the time, his dream of open country and a good horse under him, working dogs, and like-minded friends nearby had tacitly been imprinted on me.

5.

Out of the blue an editor from *Time* magazine called and asked me to write essays about visionary westerners. Two names quickly came to mind: Ray Hunt, the horse trainer and father of "natural horsemanship," and Allan Savory, the restoration ecologist who is now thought of as one of the fathers of regenerative agriculture.

In order to write the essays, I had to participate. I rode my young horse in Ray's clinics and took Allan's week-long classes in Holistic Management. Those experiences and the chance they gave me to befriend these masters changed my life and added to the nourishment my mentorship with Mike had provided. Though the two men never met, they worked the same way: they knew you would find the truth the land or the horse would tell you if you listened.

Timed grazing was the brainstorm of African-born Allan Savory. Trained as a wildlife biologist in college, he was made head of the Northern Rhodesia Game Preserve by the age of twenty-one. Later he worked with ranchers and farmers all over the world to stop overgrazing, over-resting, plowing, and burning. Grasslands were once ubiquitous. The forests that covered much of the earth grew as Pleistocene-era grazers were killed off by hungry humans. But we know even now that healthy grasslands sequester airborne carbon in their root systems, and that a dynamic grassland is dependent on the grazing animal. Allan now dedicates his life to reversing climate change. Bare ground and ineffectual rainfall—moisture that can't penetrate the soil—leads to impoverished habitats. "We are a desert-making species," he said. Degraded land is one of the root causes of climate change.

In the first of Allan's classes I attended in 1985, I learned that overgrazing is a factor of time, not numbers of animals. "I could put one donkey here and he would overgraze if left here long enough, same as five hundred head of cattle," Allan said. "When a bite is taken of the grass plant, the roots respond. Too many bites and the roots wither. That's how a desert occurs."

The classes he offered covered biological health, grazing, financial planning, restoring the water and mineral cycles, and decision-making. We learned to use testing guidelines to check the viability of each decision and its impact on all the other elements involved. In other words, we managed holistically for abundance, for the biological health of the land, for the well-being of its animals and people.

At the end of the week, I returned home and changed how I managed our small ranch. I made a land plan of deeded and leased land and divided our few thousand acres into smaller pastures using portable electric fence. Every three days I rode to the cattle and moved them into the next "paddock," thus mimicking the movement of bison on the prairie. After all, native grass co-evolved with grazing animals.

Each move had flexibility built in (a day or an hour here or there) and wiggle room for changes in weather. The cows learned quickly. As soon as I arrived at the gate to the new pasture, the animals were waiting. Fresh grass awaited them.

My new learning provided new insight. I felt I could see the ranch from above. With the land plan laid out on the kitchen table, I could make decisions based on what the day presented: on the ecosystem-wide context and the smaller wholes within the larger ones. Doing so opened my eyes to the genius of grass seed: how it can lie dormant for decades. As cattle, horses, and wildlife stimulated, aerated, and manured the open range, more native grasses appeared. With more grass came more birds, elk, deer, and water. We developed three springs by pushing stovepipe into the oozing ground and hanging the other end over a stock tank. Water for all! Wherever the excess spilled over, the whole hillside turned green. That's how we came to have a winter herd of elk. They lived on the tall, strong-stemmed Great Basin rye that began growing.

The more I saw, the more I could see: Hudson Falls became a laboratory where the quality of life for every living being was enhanced. The animals and I existed as a movable unit, and the false division between me and them quickly fell away.

"Plan, monitor, and replan": that was the daily mantra. If there was an unseasonable snow, I could move cows and calves off the mountain; if there was drought, I moved them higher. Healthy soil meant more moisture infiltrated and more carbon was stored in the ground. It was shocking to see how degraded much of Earth's land was, how fast and widely the planet was desertifying. Everywhere land was being abused. Bad range practices—continuous grazing—and commercial agriculture were to blame. "Our greatest export is topsoil," Allan said. Topsoil blows away in prevailing winds. Fallow and plowed ground destroys microbial life and absorbs more sun, making it hotter. Allan said, "Drought doesn't cause bare ground; bare ground causes drought."

Ranching with new eyes didn't mean I was exempt from the ravages of weather. On a May morning I awoke to three feet of snow. We'd just turned out the cows and young calves, but the snow was over their heads. A bellowing cow woke me: it was Spot. She had broken trail for the others and led the herd home. If she hadn't, the calves would have suffocated.

We hand-carried bales of hay from the stackyard to the cattle. We opened gates and let the horses in with the cattle to help clear the ground: horses paw through snow; cattle don't. It took all day to feed the livestock. Later I called Mike. "Your Longhorn cow saved our herd." There was a silence. Then she said, "In Wyoming it can always snow."

*

With Ray Hunt, I learned the lessons of the round corral, and there I threaded my past of dance, meditation, and horsemanship into one.

The horses were all young, two- and three-year-olds, untouched, unjaded, incomparably strong and well-bred, innocently milling around a high-sided round pen. The morning's frost held the grass stiff until a breeze lifted the sun over the ridge. There was a blue roan, a bay, and two sorrels sniffing, snorting, flicking their ears, trying to understand what was going to happen to them.

"See, they're a little troubled," Ray said, "and when their minds are troubled it shows up in their bodies." Tall, rawboned, upright in the saddle, he rode into the pen. "A horse will tell you what he understands and what he thinks about it. He's telling you all the time, but you just don't see it; you're just not willing to go that far in his direction," he said, his eyes intense as he looked at our horses. "That's okay, but you're not going to get too much back. To have a willing communication with your horse, you'll find that first you have to develop awareness and discipline within yourself so that you can have it with the horse later," he said.

Ray Hunt was raised in Idaho with a father who used draft horses to plow, plant, and cut hay. He grew up the hard way, doing farmwork to make money, but the work he was born to was cowboying. He rode the rough string—the hard-to-train horses—on a big ranch in Tuscarora, Nevada, where it was common to ride a fifty-mile circle every day. When the aspiration of owning a ranch gave way to an ardent commitment to teaching humans how to handle a horse—what used to be called "breaking colts"—he began giving five-day-long clinics in the United States, Canada, and Australia.

By the end of the first day I'd watched a green, untouched colt accept being caught, haltered, led, saddled, and ridden. By the fifth day the horse had learned the rudiments of coming to a sliding stop, turning, backing smoothly, and changing leads in an atmosphere so quiet and unhurried it was hard to believe anything had happened at all. What Ray taught had nothing to do with "breaking." When I asked him how he did all this so quickly, he said, "Oh, I just work with the horse's mind."

You only had to watch Ray in the round corral to see he didn't suffer fools gladly. "This isn't some commercial thing," he said. "I wouldn't do it. This is life. There's no rulebook on this and its damned hard to grasp because it comes from deep down inside. I've been trying my whole life and I'm still working on it. But when you do get it, pretty soon it starts coming back to you directly from the horse, and from then on, it's a continuous thing, and there's no end to what you can learn."

In the round pen that first morning, Ray leaned over from his saddle and stroked the colt's face with his big meaty hand, then did the same to another horse. Every gesture was simple and soft. He moved easily between animals, neither slow nor fast, but with an even keenness that told me everything about Ray Hunt and what he thought a proper relationship between animals and humans ought to be.

A toothpick rolled from one side of his mouth to the other. "To understand the horse, you'll find you have to work on yourself," he said matter-of-factly. Ray's lessons of the round corral involved giving, discipline, awareness, compassion, and stillness, which are exactly the Buddhist paramitas, but in western dialect and given physical form.

"How did you learn this?" I asked. "It didn't come easy," Ray explained. "I didn't just scrape off the top, and there it was. I dug and dug and tore my hair out. But I owe it to the horse to work this hard because I used to do things the true-grit way, not out of meanness, just out of ignorance. Guess I saw too many Charlie Russell paintings! I didn't know there was another way."

The true-grit way looked like this: A green horse was roped out of the remuda, led into a corral, and tied to a snubbing post, from which it struggled to get free. Then its front feet were hobbled, and the cowboy would come at it with gunnysacks, waving them at its face and all over its body. More terror and struggling followed when the cowboy threw a saddle blanket and saddle on the horse, cinched it up tight, and, jerking the horse's head to the left and grabbing some mane, stepped on fast. Around they went bucking and snorting: the harder the cowboy pulled on the halter rope, the harder the horse bucked.

That was the first lesson the human taught the horse: that it could look forward to a lifetime of domination and terror. "The horse learned that when he was afraid, to try like hell to buck the human off. With luck, the horse won. A horse can always buck," Ray said, grinning. "That's his defense mechanism, and you have to fix things up so he doesn't need it."

A horse named Hondo in Tuscarora taught a young Ray Hunt to change his ways. Hondo made it clear that Ray could be broken, but that he, the horse, could not. "Everything I know started with that horse," Ray said. "Hondo was a striking, biting, kicking, bucking tough colt who might have killed me. Day after day Hondo would say, 'Come on, try to break me and I'll break YOU instead.' I had all winter to work with him. He was my only horse out of the rough stock that year. Without him, I was afoot. It was just him and me.

"See, a horse that's had trouble can't believe a human will quit hurting him. I felt sorry for Hondo. At first he wouldn't let me in the stall without striking at me. Finally we got so I could get near him, then get on him. I'm not sayin' it was all love and kisses. Things got pretty physical, pretty western. I grabbed a lot of mane when riding him, just in case. I'd go to bed at night and think about that horse, dream about him, then go back to work with him the next day.

"Finally, I took him to California to Tom Dorrance, the man who taught me so much. Tom was a little old bowlegged cowboy, but he's the brains of it all. He can fix a horse so fast you never knew what happened. And who taught Tom? He says it was the horse.

"As soon as Tom came around me, Hondo acted like a lamb, and as soon as he left, I'd be riding a tiger again. I couldn't understand. Something was going on, but I couldn't find it."

Ray looked out over the young horses, then at me. "See, I was too forceful. The timing was good, but the mental feel of how it could be wasn't there. The horse was afraid of me. I thought I had to hurt him to get him rideable." Ray leaned forward and ran his thick hand down the neck of the horse he was riding, then sat up straight again. "I knew it wasn't right. And pretty soon, I learned to give respect in order to get respect from the horse. There's a difference between firm and forceful. And there's a spot in there, inside the horse, an opening where there is no fear or resistance, and that's what I began looking for. By the end of that year, Hondo was smooth, athletic, and kind to be around, a horse the grandkids could ride."

The sun was partway up by the time Ray had sorted the horses until just one remained in the pen. "You see, you're not working with a machine, you're working with a mind. The horse is a thinking, feeling, decision-making animal, and each one has a distinct personality. But the human always acts superior. He thinks he's smarter. He always wants to have things his way and right now. He wants to be boss. If trouble comes up, he turns it into a contest with the horse. But if you do that, watch out. You just may lose!

"What I'm talking about is not dominating with fear, but more like dancing with a partner. It's all balance, timing, rhythm, the kind of dancing where your body and his body become one."

In the morning, steam rose from the creek as Ray worked a filly in the round corral. She had no saddle or bridle, no halter. She traveled smoothly, trotting first one way, then the other. Ray wanted her to loosen up first, then turn off the rail, stop, and come to him. "I'm doing some things now that will let the horse accept being caught. It's awful hard to ride them if you can't catch them first!" he said, grinning. He swung his lariat and she broke into a lope, stopped, and when she tried to turn away instead of coming into the center, he insisted she keep moving. "See, I'm making the wrong thing hard and the right thing easy." The horse kept running around, but when she found out nothing was going to happen to her, she came to a stop.

"A horse gets sure and unsure, scared and bold; she says 'Maybe, all right, I don't know.' But I'm going to show her that things can be all right." She panicked again and stuck her nose over the top of the pen as if to jump out, but Ray kept her moving. "All I do is operate the life in the body—through the legs to the feet to the mind. Pretty soon she'll come off that rail; she'll turn loose and stop trying to escape," he said, watching calmly, and just then, the filly stopped, moved her hindquarters around, and pointed her ears at Ray.

"There's a change," he said in a low voice, and kept watching. She trotted around at the rail but began to relax. When she stopped, he walked over to her and stroked her head once with his hand. Then she darted away. "That's okay. I'm not going to make her stay. She's still afraid I might hurt her, and she needs to know she can escape. She's telling me she's not quite ready for anything more."

By afternoon, Ray had a lariat loose around her neck and with it, he bent her neck to one side and the other, moving her hindquarters until her front feet followed through, then led her forward a few steps and back. "You see, a horse is much stronger than me, but if I prepare her for dancing instead of fighting, I may survive."

Next came the saddle blanket and saddle. He let her see it, sniff it, and he put it on her back with utmost gentleness. "See, I don't sneak my outfit on the horse; I put it on respectfully." The horse was turned loose in the pen and she bucked a few times, then lunged, struck, and bucked again, snorting each time. Ray watched calmly. "There's a change," he said as she finally walked to him, working her mouth—a sign of relaxation.

"I give them a place where they can come to me. They see it in my body." The horse stopped halfway and thought about leaving. Ray watched her but made no move to catch her. "That's good with the mind. Now here come the feet," he said, and she "turned loose," walked over to him and stood quietly. "See, she had to check things out first. That doesn't make her wrong. You don't punish a child for being afraid. She already knows my mind and I try to understand hers. She knows I'm a friend."

Soon all the colts were ready to ride. One by one they were

saddled and ridden with no bridle—just the halter and a lead rope. "It keeps you humble to ride a horse with nothing on their head," Ray said. "It forbids you to try to control the horse, and the horse feels that—boy, does he feel it—and that's the beginning of trust."

A young man stepped into the stirrup. His sorrel colt flinched, so he stepped off, waited, and tried again. "He likes that. He wants to know everything is going to be all right and I bet you do too," Ray said, grinning. The rider threw his leg over. "There you go. Good luck!" Ray said, and just then, the colt blew up, bucking and snorting. Ray yelled over the commotion: "If you pull on that lead rope, you'll be teaching that young innocent horse how to buck every time you get on." The horse bucked a few more times, then settled. "See, he found out it's easier to be quiet. Get them used to you and they'll accept you. You want to be just like their mane and tail."

When all the young horses had riders on their backs, Ray worked on what Buddhists call *dana* and *virya*—the two paramitas that concern giving and vitality, and had the riders ask their horses to move. "Have a lively feeling in your body and they'll get one in theirs too."

The next day the horses were bridled with snaffle bits. "The reins aren't for control but for sending messages," Ray reminded the riders. He worked on *dhyana* and *upaya*—concentration and skillful means. "The reins should feel like silk in your hands. There should be a float in them. You should feel weightless." As they rode in the big arena at a walk, trot, and lope, Ray said, "See how little you can do. Bring the horse to a walk without using the reins. It should be in your body. See how slow and soft life can be without letting things die."

Later that day I trotted by Ray on my tall sorrel horse and he yelled at me: "Pick up a feel and speed him up. Don't sit there like a gut-shot bird!" Laughter. "Your legs should be doing everything," he said, and trotted out to demonstrate: his horse turned one way, then the other, slowed down and sped up and Ray hardly moved in the saddle. But when he loped his horse around in a circle, the mare clamped her tail, ready to buck, and Ray grabbed a handful of mane—not the reins—and gave her something new to do: ten figure eights one way, then the other. Then he stopped her in the middle of the arena. "See, she had something on her mind, so I took those ideas and turned them into something else without punishing her."

Things in the arena became slightly chaotic. He asked us to count cadence—tell him where each foot was as we trotted by. "It's one mind and one body," Ray yelled out. My horse and I were behind Ray, where he couldn't see me, when something happened: my reins were loose and each of the horse's feet seemed to drive up inside me like pistons. We floated. It felt like nothing else I'd ever experienced. Ray's back was to me, but he knew: "That's right, Gretel." As Slim and I came around in front of Ray, still moving smoothly, he turned, looked at the horse, then at me, and smiled.

Last day. Ray sat his horse as we gathered around. His voice was raspy from dust and fatigue. "The horse is a mirror," he began, rubbing his hand down the mare's neck. "It goes deep into the body. When I see your horse, I see you too. It shows me everything you are, everything about the horse. I'm trying to save the horse's life and your life too. The human is so good at war. He knows how to fight. But making peace, boy, that's the hardest thing for a human. But once you start giving, you won't believe how much you get back."

He paused and looked around. "You have to be on the spot

every moment with a horse because that's where he is. Fix it up and let it work. It should be like silk all the way. It's hard to teach what I've been talking about all week because the first thing you need to know is the last thing you'll learn. But I can tell you this: when you get to square ten, all of square one will be in it."

6.

I had been trying to learn how to run a ranch, enhance the operations of photosynthesis, and find true harmony with horses. The ranch thrived but the marriage came apart. There were serious problems—both mine and his—that we failed to iron out.

In May 1991, the night before I left for months of work outside the country—first to Nunavut, Canada, on assignment for *Harper's Magazine,* then to London to write a poem cycle for a choreographer—I was served with divorce papers. I asked the lawyer for a continuance and traveled on.

A life can change so suddenly and completely, nothing about it is recognizable. In August of that year I was hit by lightning. During a break from rehearsals before opening night at London's South Bank Centre, I'd returned to the ranch in Wyoming for a week's respite. The second night I was there, I was hit by lightning.

A phone call from old friends, Sonia and Robert, saved my life. I had made my way back to the house after being unconscious for hours, but my throat and left side were paralyzed. I couldn't talk and could make only a horrible squeal. They called 911 and after an hour or so, two first responders arrived. I was taken to a badly staffed rural hospital and released the next day:

barefoot, with unwashed cuts all over my body, dazed. After several attempts to find good care at three regional hospitals, I retreated to the ranch and that night, alone and unable to stay conscious for long, I saw death: there were black figures in the corners of the room coming for me.

I made an unusual decision: to call my parents and ask for help. My father asked if I could get on a flight to California. I said I didn't think so. He came for me in a friend's plane and I was flown to the hospital in my childhood home, Santa Barbara. Back to square one, as Ray Hunt would say.

On the cardiac care unit, a nurse said, "You can relax now. We're not going to let you die." My cardiologist read all my books while I was in the hospital. "Now I know how to take care of you," he said. When the electrochemical messages to my heart that told it to beat slowed and, at one point, stopped, I was revived and later, put on Norpace to keep my heart ticking, but staying conscious was difficult. "You can do anything you want," my cardiologist said with a smile, "as long as you do it lying down, and that doesn't include sex, alcohol, hot tubs, or anything that might require you to stand up too long."

Ten days later I was released to my parents' care. They were seventy-five and I couldn't keep up with them. When my father walked me to the mailbox yards away, I clung to his arm and barely made it back. While I was resting, my mother stroked the top of my hand. I cried; I was their baby again and I needed them to survive.

The lightning had caused a brain stem injury, and my sympathetic nervous system was fried. I could barely process a thought. Deaths had occurred—not mine, but that of my cowboying life, my marriage, and deep friendships. They were wiped out, and I was back living in the town from which I'd been dispatched in order to "make something of myself." I was forty-five years old. Enormous self-protective blanks installed themselves in my brain: the lightning blank, the ranch blank, the divorce blank.

An old friend found a house for me. It belonged to the painter Ken Noland. Spare and simple, it was built on stilts in the sand. Every wave at high tide shook the house. Every wave entered me, all timbre and tempo, telling my heart to beat. Evenings, I forced myself to watch the sun sink below the ocean's horizon and made a pact with myself to never miss a sunset. In other words, I had to stay alive to see it. The cowboys in Wyoming always said at the end of the year, "Let's hope we make it to green grass." That was my motto too, but for a long time, I wasn't sure I would.

My health improved, though more slowly than I'd thought possible. It took three years. The Spanish-speaking checkout women at my mother's grocery store prayed for me, as did my parents' housekeeper, Pauline. That was good enough for me. I sought out no physical therapy. Sammy, my dog, had been flown to me. A friend in the movie business organized a car service to pick him up at LAX and delivered him to me at the beach. Sam was my mirror, my only companion. He watched, curious, as I stood on my head, legs propped against the wall, to get blood and oxygen flowing to my brain; then we'd wander aimlessly on the beach as far and as often as we could. Sometimes a few yards. Later, miles.

I slept. The fatigue was bone-deep. I listened to my body: what to eat and how often to sleep and when to walk. My kelpie-heeler cross and I made beach friends: Hillary Hauser and her schipperkes, the Australians Kate and Clyde Packer and their German shepherds. Every Wednesday was barbecue night: one "barbie" was set up for the dogs and one for the humans. I was happy in my shambling life, too unwell to notice how I must have seemed. That was the good part: I didn't know how decrepit I was, so I didn't care.

The medication that instructed my heart to beat made it difficult to read. Instead, I listened avidly to chamber music, the string quartets of Beethoven, Bartók, Shostakovich, and Kevin Volans. I was convinced that its cerebral intimacy would repair my brain.

My parents were more concerned than I was about my invalidity. I'd faced death and won (with their help) and felt confident my body knew what it was doing. When the neurologist announced that brain cells don't regenerate, I refused to believe him. How could that be possible when all the other cells in a body grow back? The true setbacks were emotional: the divorce was finalized; I lost the ranch and my investment in it. I asked for my equal share and didn't get it. Instead, I received two horses and one dog. Two years later, my books and clothes were sent COD by truck to California.

One morning, in the third year of convalescence, I felt my ch'i rise, and sat up in bed laughing. Sammy put his paw on my arm and growled with delight. It was not so much that I wanted to live—I simply found myself fully alive. I called my cardiologist. The nurses in his office cheered: "Get down here, we want to take your vitals!"

For the first time, my lagging blood pressure was almost normal; I no longer felt as if a black sky was lowering down the middle of my forehead, or that I was going to faint if I stood. My ch'i kept announcing itself as the old vitality returned, so I set out, less strong and more vulnerable to fatigue, stress, and heat, yet quite jolly between agonizing bouts of ranch grief. I began to consider my decrepitude a blessing, an unexpected opening into a deepened life. My sister said I was "nicer." A visiting monk from Japan said, "You know everything about being strong. Now you have to learn about being weak. Otherwise you will be out of balance."

Nights, I longed for the Wyoming mountains where I could

ride and walk in total privacy. Then I found out that our end-of-the-road ranch had been sold without my knowledge or consent, and there was nothing I could do about it. As the effects of the lightning eased off and things slipped away, other delights took hold. Like a volcano that destroys part of an island but makes new islands in the process, I was eager to investigate unexplored territory.

In those early months after being released from the hospital, Blaine, my cardiologist, said it would be best if I stayed in the area near good medical care. No one knew what the long-term effects of a lightning strike might be. He loaned me money for a down payment and I bought a hundred-acre parcel on the Hollister Ranch, where I had ridden when I was a teenager—the Hollisters were friends of my parents. There, I could have some semblance of a ranch life with dogs and horses.

The parcel came with an abandoned, half-built house and a habitable guesthouse perched on a hill, half a mile from the ocean. Bill, an eccentric surfer-contractor, finished the buildings. His motto: "Surf up, hammers down." I had exactly $40,000 to spend. We scavenged wood from the beach, and after historic El Niño floods, Bill bargained for the goods from downtown stores that had flooded: bathtub, toilet, sinks, even a dishwasher. Quite exciting for someone who had lived in an uninsulated house in the mountains of Wyoming where kerosene lamps and melting snow on the wood cookstove for water were the norm.

Bill and his brother made a cement floor divided by pieces of driftwood. He set a bathtub down into the thousand-square-foot deck. The kitchen counter was a raw slab of unpolished Carrara marble he picked up for a few hundred dollars. Large driftwood logs that had rolled onto the beach were stuccoed into the drywall corners.

The house became a sanctuary of sorts from my sins of confusion. The solace I'd found in Wyoming after David's death was no longer available to me, but I'd found a place near what the Chumash Indians call the Western Gate—the place where the dead go—with its wild nocturnal winds and miles of beach backed by chaparral-covered mountains and enough space to walk and ride where no one would bother me.

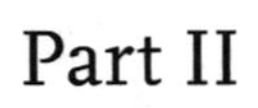

Part II

7.

In 1993 an unforeseen vitality pushed me back into the world. With no ranch holding me, I plunged: nothing more to lose. Not that I went out seeking excitement; rather, it kept coming to me and I didn't hesitate to welcome its arrival.

A chance meeting with the editor of *Islands* magazine resulted in a chance to go to Greenland, and off I went with no real idea of where I was going. On the way I met a young couple from Uummannaq, a village on Greenland's west coast, and they drew me north for the summer, and for part of the next dark winter.

The next spring, when I returned, the sea ice failed to come in. We had planned to travel up the Melville coast by dogsled, a route that Knud Rasmussen and Peter Freuchen regularly took in the early 1900s. Little did we know that such a trip would soon be impossible, that the Davis Strait would not freeze again.

Following my friends' advice, I went north, to the two northernmost villages in the world: Qaanaaq and Siorapaluk, traveling with the elite hunters of the far north: Jens Danielsen, Mamarut Kristensen, and their extended families, about which I wrote in my book *This Cold Heaven*. I returned often—almost every spring—and for the next twenty years they allowed me to accompany them. Against all odds, these families had chosen to live a traditional ice-age life, taking what they needed from the modern world and insisting on the old ways. When I asked why, they said, "Dogsleds and skin clothing work better."

Greenland subsistence hunters are soft-spoken and savvy, as elegant in the way they make decisions and handle a dogsled on the ice as when they manage the social well-being of their villages. The dogsleds are long and sturdy—fourteen by four feet—and are pulled by fifteen to twenty dogs harnessed in a fan hitch and spread out across the ice because there are no trees.

In those days before the climate heated up, the ice came in mid-September and for nine months it was possible to travel and hunt on the frozen sea. By Midsummer Night, the white floor of the world quickly disappeared, and bright all-day, all-night light shone down. By October 24, the sky goes dark and the sun doesn't reappear until late February. In March the spring hunt for food urgently resumes.

In Greenland there is no ownership of land. What you own is your house, your dogs, your sleds and kayaks. Everyone is fed. It is a food-sharing society in which the whole population is kept in mind—the widows, elderly, infirm, and ill are always taken care of. Jens said, "We weren't born to buy and sell, but to be out on the ice with our families."

Ice-adapted people have everything to teach us: they have a survivor's toolbox of self-discipline, patience, and precision. They understand transience, chance, and change. Adaptation for them is a daily flexing of mind and muscle. Animal-human transformations and spirits are the norm. Utterly practical, conscientious, observant, clever, and wise, Jens and his extended family knew that the individual mattered less than the group, that the animals understood what humans said, that there are consequences for every human action. They had survived for five thousand years at high latitudes by depending on the group's hunting together, helping one another, and sharing food. Jens's wife, Ilaitsuk, said, "We all have our differences. We just keep ours inside."

Sea ice—ice that freezes ocean water and floats on top (as distinguished from glacier ice)—is a Greenlander's highway. Usually ten feet thick, it is the icy platform on which Arctic animals, including humans, travel, rest, bear young, and find food. In 1997, the dogsled I traveled on with Jens and his friend Nils almost went through a hole in the ice. Jens retrieved the fifteen dogs from the water, and we were able to continue on. But he was shocked. "This shouldn't happen. Something's wrong. Please find out why." After, I dedicated my days to educating myself about why the Arctic climate was changing and how it would affect the entire world.

Albedo (from *albus*, Latin for "white") is essential. Without it, the planet cannot reflect the immense solar heat it receives back into space, and thus, keep the lower latitudes temperate. The less albedo, the more global heat. Every time I returned to Greenland, I witnessed what James Lovelock called "Earth's morbid fever" and the demise of my beloved friends' cold heaven.

Jens's dogs had gone through the sea ice in 1997. By the year 2000, Arctic scientists who studied ice around the top of the world were saying that Greenland's northern district was part of a "rotten ice regime." Instead of nine months of dependable sea ice, there were, at best, only two or three. Ten-foot-thick ice had dwindled to a mere seven inches.

I could never have imagined the extent of the changes taking place. I'd gone to Greenland to celebrate Inuit culture and life on the ice. Ironically, my new life on a dogsled in Greenland with Jens, Mamarut, Gedeon, Mikele, and their wives contained in it the dark seeds of mortality and extinction. Hard times would soon come.

In 2002, I was invited to be part of the Greenland delegation to the Inuit Circumpolar Conference held in Utqiagvik (formerly Barrow), Alaska. It was there that one of the delegates, an anguished hunter from Qaanaaq, said to me: "This is the first time I've been told that we are losing ice. What is it that I'm to tell my children? Everything I know has to do with ice. Now I will have nothing to teach them."

Climate is culture: as soon as the ice in the Arctic began to disappear, so did the lifeways of Greenland. Arctic people had traveled over the land bridge across Arctic Alaska to Nunavut and finally to Greenland five thousand years ago. They were part of an ongoing Arctic culture that was at least thirty thousand years old. Though the polar traverse was long, the language, lifeways, watercraft, clothing, and social mores remained robust enough to endure. A single culture spans the entire top of the world, from Siberia to eastern Greenland.

Now the youngest members of that culture were holding on to the stray ends of a death spiral. They were being sent to south Greenland to attend college or trade school. Instead of living the hunting life on dogsleds, they were becoming electricians, cooks, helicopter pilots, nurses, or teachers. To drive a dogsled and join the elite marine mammal hunters had once been tantamount to becoming a national treasure. Now it was no longer an option.

8.

Between another Greenland spring hunting season and midsummer, when the ice floor vanished and the world in front of Qaanaaq and Siorapaluk turned liquid again, I returned home to rest and write in my rustic, Big Sur–type house forty miles north of Montecito, where my parents had lived since the start of World War II.

The dogs exploded with happiness when I returned. We ran up and down the beach as the tide came in, racing with sanderlings up and back, then in a straight line toward Point Conception.

Persistent drought had taken hold there, as well as in the Middle East, southern Europe, and Australia. As the Arctic sea ice thinned, there was less reflectivity to drive away solar heat. I knew the months and years were on a trajectory toward getting warmer, then hotter and drier, but for the time being I enjoyed the winter days.

Spring arrived a month early, in February, with blue ceanothus and hillsides of mariposa lilies, Indian paintbrush, and owl's clover. "The world is its own magic," the teacher Shunryu Suzuki said. Feral pigs gamboled with newborn calves in the meadow below, and the dogs and I went there just to watch.

Bill, the contractor, installed a Dutch door on one side of

the house, and the horses put their heads through the open top door and watched me write. The dogs lay under the desk. As soon as they heard my computer go off, they would leap to their feet, begging for action. Action meant long steep walks in the hills or on the beaches.

From almost the first day in my new house, I'd made friends with John and Linda Kiewit. They were serious readers and world travelers. We often drove to the far end of the ranch to visit Jane Hollister, the daughter of the ranch's founder, and her husband. Joe was a journalist who had accompanied Mao on his Long Walk in China. Jane bore their first child there. Later they studied with Jung in Switzerland, came home and practiced therapy in San Francisco, then retired to Jane's home ground, Hollister Ranch.

From their house we could see Government Point, Point Conception, and Point Arguello, where California's elbow bent and turned the state north. John and Linda and I would drive halfway up the mountain, turn the music up, get out, and dance in the dirt track. Strong winds grabbed us; the peppermint scent of eucalyptus spread everywhere. The sea glinted as if watching us, winking at our frivolity. With his large-format camera, John photographed us in light so translucent we looked permanently overexposed. We stopped to admire the twisting branches of manzanita trees that grew on the ridgetop, wind-tortured and smooth-skinned, overlooking the ocean miles below.

That same week a canyon wren adopted me. He used the Dutch door too. His cascading song woke me each morning and reminded me of a bearded seal's mating song. He had a curved beak, a short tail, and a pale breast, and spent days digging out insects from between the wood logs of my house. He slept on a four-by-four beam that spanned my bedroom.

The day the wren brought a girlfriend home, I felt jealous—that's how absurdly solitary I was. She was difficult and I didn't like her. I'd laid dust balls, dog hair, and thin oak twigs on the hearth for nesting material and he worked hard constructing a nest. She'd watch him without helping, inspect it, and time after time, tear it down and wait for him to build another one.

"I'd never treat you that way!" I yelled to him as he swooped through the open door. Finally, she approved of the nest he built on the high beam in the bedroom and she laid four eggs there. From bed I watched through binoculars—not that the room was big, but I wanted an intimate view. *Catherpes mexicanus:* the canyon wren had a white bib and a slightly curved beak, and its song is "a gushing cadence of clear curved notes tripping down the scale." Who can top Roger Tory Peterson's description?

Every morning the wrens woke me, and I'd check the nest. One day I heard peeping and saw chicks pecking their way out of their shells. Heads popped up. A week later they were all over the house, fluttering and landing and taking off again. They used the hallway for touch-and-goes. From bedroom to kitchen there was bird shit everywhere. Bill built a wonderful birdhouse with a sign that read EARLY MORNINGS AT THE BIRD CAFÉ. I filled it with birdseed and the mice ate it all.

Finally, I papered the floor with the Sunday *New York Times*. The chicks landed on my head and arms while I was typing as the parents dashed in and out with a fly or mosquito clasped in their beaks. By now the four sliding glass doors that faced the deck were open all the time. One rat came in and out, but he respected me: I could make him leave if I yelled loudly enough.

The day the chicks fledged, it was excessively windy. One chick was blown off the deck and died. I buried him by the Scotch pine I'd planted years before. The wind-hiss through pine needles was consoling, but I was sad for the bird and had begun to feel a bird's spring restlessness, longing to be on the move again. But I had obligations.

My parents had saved my life after I was hit by lightning,

and I'd made a pact with them to help in their last years. My mother died first in her usual way of doing things: elegantly, privately, and with no complaints. She'd quit smoking that way and decided to die with the same resolve. Of course I didn't know that was what she'd planned. She didn't tell me. I had been at their house and wanted to make dinner for them and spend the night, but she insisted I go home. Her last words to me were: "Go feed your animals. You need a haircut. Get your house cleaned."

That night my mother told the housekeeper that she was "tired of living this way." Not sure what she meant. She was in her eighties and had found the "golden years" less than enticing. More she'd never told me, or was it just death knocking? That night she ate two ice cream bonbons and drank a glass of champagne. Then she went to bed in the guest room, which puzzled my father since they had never spent a night apart.

Dying demands privacy. I'd seen it in animals. The last thing they want is someone staring at them and fussing. She died at dawn. It was the night before New Year's Eve and her long evening gown was laid out over the back of a chair.

That morning my father called. In a frail voice he said: "I can't seem to wake up your mother." I knew she was dead. It took an hour to get to their house, and I went straight into the guest room. She was lying on her back, her eyes closed. She looked as beautiful as ever. Carefully, I removed the gold wire with the diamond she wore around her neck. To lose your mother is to lose the deepest aspects of your history. Only she really knew me, and in ways I didn't know myself.

Days later, when they lowered her coffin into the grave, the hair-tearing grief finally came. Being buried was too claustrophobic—having dirt thrown into your face. I wanted to pull her out and carry her to the windiest point above the sea.

Eventually my grief eased into a month-long trance as if still attached to the umbilical, but weightless somehow, and drifting.

My father died the next year on tax day. Perfect, I thought. He was furious with himself for having smoked all his life, since he was otherwise perfectly healthy. We had round-the-clock male nurses with whom he spoke Spanish. I was on deadline for a small book, so I set up a folding table by his bed and wrote while he slept. He'd wake up from time to time and smile: "Aren't you finished yet?" We'd laugh and he'd sleep again.

Every night at six o'clock I made his usual martini: vodka, lemon peel, ice, and a hint of vermouth. The last month of his life he couldn't really drink it but asked that I pass the glass under his nose. I hoped the scent of vodka assuaged what must have been the terror of not being able to breathe.

The day his physician called and asked me if my father wanted morphine, I faced a dilemma. Morphine would depress his respiratory system and result in death. I was being asked to kill my own father. Reluctantly I entered his room, sat beside him, and finally asked the question. He looked hard at me, then with a faint smile, shook his head no. I ran to the bathroom and sobbed in relief.

That night I stayed and told the nurses to leave. I could take care of him. Sometime after midnight he woke unable to breathe properly and had a bout of diarrhea, so I hauled him half over my back to the toilet, then sat on a chair in front of him and started what he called the "breathing machine." Tube in mouth, he sucked in medicated, filtered air and I sat opposite, flushing the toilet. Once he'd caught his breath, he started laughing and I did too. "Father-daughter quality time," I said. He nodded. Still laughing, I carried him back to bed.

In the morning I found him lying on his back with his hands crossed over his chest. "What are you doing?!" I asked, slightly exasperated. He said, "Practicing dying," and gave me another sly smile. That same week he asked what I was going to do when he was gone, and I said, "Go back to Wyoming, buy land, have a cabin built." He smiled and said, "Good."

I wanted a ranch but knew better than to think I would be happy ranching alone. A small place at the foot of big mountains would be sufficient. To stay in Montecito was anathema. His bon vivant life didn't interest me, even if I'd had the money to support it. I needed to strip away anything that impeded the feel of the earth, anything that obscured direct access.

He took his last breath as dolphins swam by. After, the nurse, Luis, and I sat with him all afternoon, telling him what we were seeing outside and reminding him not to be frightened. The sea sparkled and a few sailboats glided by. My father had been an ardent sailor, so I described the boats: a ketch with an aft jib, a sloop with a huge mainsail, another one with a luffing genoa.

I felt strangely euphoric, the way I had felt watching calves being born. When the room grew dark, we called the undertakers. Later, Luis and I walked the beach. There was a moon and the sand seemed bright. Luis said, "Don't be sad. Your father is just being recycled." But after, alone at my house forty miles up the coast, desolation filled the rooms. I gasped, bent over suffocating, then stood straight and inhaled fresh air.

9.

Since the day I'd met John and Linda Kiewit seven years earlier, they always had dinner waiting for me whenever I returned from one of my trips. We'd stand out on the terrace and watch the sun go down, then go inside and eat. There were piles of books on tables—the old ones we were all rereading, and science, natural history, poetry, and fiction side by side. There was music in the house and sometimes we left off talking or reading and the three of us danced for a few minutes, or else we cooked, thumbing through old issues of *Gourmet*. The ocean was part of our friendship. From above, on the hill where their house was perched, the Pacific was a sensuous, undulating mass with sea otters lying on their backs cracking open mussels. At low tide the tide pools shimmered. We often walked the beach for ten miles.

One afternoon Linda called and asked me to come over. She wanted to talk. She had been fighting breast cancer for ten years and it had returned. Her cancer had spread to her liver and she knew her time was limited. John poured wine and we sat side by side on the couch. Linda sat opposite. A strange smile came over her face before she spoke. Then she asked if John and I would be willing to marry each other when she was gone. I looked at her, stunned, and unconsciously put my hand over my mouth. What was I supposed to say?

"You are such good pals, I know you would be happy," she said. John's eyes widened. He stood, uncorked a bottle of Sancerre, and poured three glasses. We drank in silence, then wandered out onto the terrace and watched the rising and lowering kelp-matted sea.

That night, alone in my house only two miles from theirs, I tried to configure past and future. Was she lying beside John trying to imagine being dead? Was she lying beside him thinking of me by his side? In saying goodbye this way, I saw she wanted things to be orderly and true, and the happiness she'd had with John to be his and mine. But did John want me? Did I want him? Did he want his future to be mapped out this way? She had been the love of his life and I wondered if any of these plans would survive his grief.

John called a few days after the "proposal" and we agreed to meet. When he came into the house, he looked as if he'd been dipped in salt and laid out in the sun. His shirt was pale blue. He wore no shoes. I led him by the hand to the middle of the room, then turned to face him, questioning the next move.

There was no hesitation. Our passion for each other surprised us, as if something had been torn away. We made love urgently because time mattered. For Linda, time was running out and for us it was just beginning. John came back to the house over the months, not to prove a point but because we found we cared for each other more than we knew.

When Linda died, I gave the eulogy, and after, John retreated from me. Grieving takes on its own life in which anything can happen. Was Linda's ghost swarming him? Was he hiding or running? And why didn't he come to me? Then I heard he'd gone off with another woman. When she rejected him, he found doctors to prescribe all kinds of pills, which he mixed with drink and with wandering.

His parents called me to see if I could help, and together we planned to take him away. I rented a house in the Shields Valley in Montana where all four of us could live for a month. I finally broke the silence and told John the plan and was surprised when he eagerly agreed. But when the time came to leave, he refused and vanished again.

Later I heard that mutual friends had set him up with someone I knew. I was shocked and angry. No one was aware of the pact we'd made with Linda. But wasn't our ambivalence a betrayal of her trust? We owed it to her to try. But I felt powerless to change things. The promise we'd made became an impediment rather than a way forward. He ran from it. I did nothing.

One night, John did come to my house, but he was out of his head, afraid and shaking and crying, so I put him in a sleeping bag on the couch, zipped it up, held him close, then slept by him on the floor all night. By morning he seemed okay. We had breakfast. He said he wanted to swim, and we went to the beach. He shed his clothes and ran naked into the surf. I walked into the water worried that he wouldn't come back, but just then he turned around, shouted hello, and swam back to me.

We walked a mile or two. I didn't know if I should broach the subject of getting together, so I didn't. I didn't want to scare him. But with things as they were, there was no way to sort out our feelings. We kissed goodbye and he went home. I never saw him again.

His new girlfriend had decided he should go into rehab in Arizona. He protested. I don't know what kind of push-and-pull, blame-and-shame game existed between them, or why he acceded to her wishes, but he told her that if she made him go, he'd commit suicide. Of course, I knew nothing of it at the time or I would have intervened. Mistakenly I thought that grief had to play itself out, that eventually he would knock on my door.

Instead of the smooth pathway Linda hoped she had laid out for us—incongruous as it sounded—a terrifying morass had opened.

In relationships, my old friend Pema Chödrön said, "Not too tight, not too loose." But I miscalculated. From rehab John called his girlfriend. She said she was too busy to talk: she was having a massage. He went to lunch with the others, then excused himself to go to the bathroom. While they waited for him to return, he hanged himself. Linda had been dead for a year. Now he was gone too.

With both friends absent, my enthusiasm about living full time at Hollister Ranch began to fade. I often stood on the sand where I'd last seen John and revisited my behavior, all the things I should have done differently. I learned to pay closer attention, to be less laissez-faire about friendships and love. My seeming nonchalance had been, perhaps, a defense, lest he found he hadn't wanted me at all. I wasn't sure whether I wanted him either. How could we know? But it would have been better to hold the reins a little tighter, exude more trust, and to have put those lessons into practice sooner. Furious at him, furious at myself, I miss him still.

10.

Just as I wondered whether I would ever be held to a place or a person again, I met Al and Russ Vail and Al's daughter, Nita. Her great-grandparents had ranched in Arizona; her grandparents formed a partnership with another family, the Vickers, and in the 1880s bought Santa Rosa Island off the coast of Santa Barbara. It stayed in the family for over one hundred years.

The night I met Nita, her father invited me to come out to the island and help move cattle. The following week Nita met me at General Aviation with our saddles, bridles, chaps, and dogs. During the forty-minute flight across the channel, the pilot dipped low over a blue whale whose back was longer than the plane.

When the island came into view, it looked like the head of a pin. As we approached, the east and west ends spread out into wide grasslands and a few narrow strands of sand showed. Once the Channel Islands were linked; now the passages between islands were rough and shark-glutted in a churning gyre. We flew in over Carrington Point, buzzed the foreman's house to let him know we were coming, turned, and made our approach. The pilot was sweating. A few yards from touchdown on the narrow grass strip, a crosswind tossed us sideways; then we bumped over a barranca, sashaying. “Ehaaa,” Nita yelled, laughing. Home.

Bill Wallace, the foreman, and his wife, Meredith, greeted us in the island’s rusted pickup with Sissy, their black Lab, firmly planted on the seat between them. Raised on ranches in northern Nevada, Bill was short and wiry. He wore his hat cocked to one side and puffed on a cigarette. He said his face was a map of the rough country where he’d grown up. “My father and grandfather cowboyed around Austin and Battle Mountain, Nevada. Everyone on both sides always worked ranches. I was made to be a cowboy; there’s nothing else to it.”

Bill came to the island as a cowboy when he was twenty, in 1948, and later became foreman. It was fifty-three thousand acres. He rarely left. “Once you get out here, it’s damned hard to find a reason to ever go back. I go once a year to the dentist and that’s too damned often.”

It was September and the air was hot and dry. Bill talked baby talk to Sissy as we bumped down the dirt track to headquarters at Bechers Bay. “The cattle are fallin’ off,” he told Nita. “Haven’t had a damned drop of rain since January twenty-sixth. That makes it about eight months. These cattle are just a bundle of bones this year. They’re walking in each other’s shadows.”

Ahead were the buildings: the 1865 L-shaped headquarters, the cookhouse and bunkhouse, Bill and Meredith’s house, two faded red barns, sorting corrals and alleyways, and beyond, the long pier where the supply boat called the *Vaquero* was docked.

Nita was in her forties, her natural platinum hair cut short with bangs. Charismatic, with a wide flashing smile, she lived and worked in Sacramento and came to the island when she could. We unpacked in our upstairs rooms, fed the dogs, and walked to the cookhouse. On the big commercial stove were frijoles, arroz, tortillas, lechuga con tomate, and carne—ranch steaks cooked to order. Outside the wind howled. The talk was of ghosts—Howard, the cook who burned up in the old cook-house, and others who committed suicide. "The damned wind got to them," Bill commented as the windows shook.

By afternoon, the gale warnings had gone up. It was the time of year when an inventory of all the cattle was made. They are brought in, treated for parasites, then turned out and moved to outer pastures. Nita and I were expected to be at Bill's house for coffee every morning at four a.m., fog or no fog, rain or shine. Sometimes we sat in his kitchen for hours, but as soon as the weather cleared, we'd saddle up and ride out at a trot—sometimes for hours in the island's rough country—looking for cattle.

The island was mostly mountain: vertiginous slopes, deep canyons, and sharp cliffs at the water. From up top we could see down to the modest human dwellings. On the lee side of the island it was grass all the way down to kelp-covered sand and turquoise water.

The cowboys were Jesús, Pancho, Arturo, José, and sometimes Beto—a congenial group of vaqueros from Mexico. Some had been on the island most of their lives. José kept the kitchen stocked and did much of the cooking. Jesús was Al Vail's best friend and a natural horseman; Arturo and Pancho were good cowboys. Cattle and island supplies were ferried out from Santa Barbara in the Vails' converted livestock boat, the *Vaquero II,* a sixty-five-foot wood-hulled freight ship able to carry fifty tons of livestock, built for the Vails and captained by Russ Collins.

Calves and yearlings were brought to the island from November through February when, on another part of the ranch, the stallion was put in with the broodmares. In May, June, and July, cattle were gathered and shipped before the Fourth of July. In August, young colts were trained. September was fall inventory, when all the cattle were gathered and weighed. October, the boat was put in dry dock, scraped, repaired, and repainted.

November, it began all over again. As Al Vail said, "All the same, but never the same year to year."

Island life generated a measured intimacy. I first went there in 1996 and continued to go until September 1998. I thought of our life out there as music: vigorous but contained, resplendent but played as an adagio. We rode together, cooked together, ate together, and slept in the ranch's modest dwellings with the heaving respiration of the ocean coming into our ears. Seals barked, turquoise waves gnawed at sandstone cliffs. Streams seeped down hidden canyons and sea lions commanded the outer rocks. Some days we found cattle on beaches eating seaweed.

Wind was our constant companion, and its attendant, the fog. For some, the surrounding ocean might have been a moat for ego or else a prison, but here, it stood for a rare kind of liberation. Al and Russ set the tone: modesty and a quiet way of doing things reigned, and it was tacitly understood that the beauty of the island was bigger than any of us. When we gathered cattle and trailed them to the sorting corrals, there was no whooping and hollering. Bill and Al both said, "Out here, we walk the cattle home."

The fall of 1996 was dry. Drought wasn't unusual. There were ancient Chumash Indian stories of hundred-year droughts on the island. Mornings, Nita's alarm went off at four. Black Mountain was flying apart, and fog scraps put it back together. José rode up the steep hill in the dark to gather horses. A short time later, we heard them pounding down and funneling into the huge corral.

In the dark at the barn—a three-sided structure with a long row of saddles—we waited for sunrise. The morning star was visible. "Is that Venus?" I asked Bill. "Hell yes, on the damned half shell!" he said, and turning from the wind, cupped his hands to light a cigarette. Behind our backs fog pumped over the island's highest mountain, crawled down to grassy headlands, weaving between horses, erasing buildings, corrals, and the pier. Daylight, when it came, looked like a void.

"Let's go get 'em," Bill finally announced. We checked our cinches, pulled our caps down tight, and headed out. Pancho opened the corral gate as Bill and the rest of us rode through. All month we gathered cattle, up and over mountains from Pocket Field, Lepe, Arlington, China Camp, Wreck Canyon, Water Canyon. At the corrals they were sorted, doctored, and turned out again.

By the beginning of the new year, 1997, one of the most powerful El Niño events in recorded history had slammed into California and Santa Rosa Island was cut off. I tried to fly out, but the airport was closed. Lower Santa Barbara flooded. My own roads became impassable. Bill and Meredith were on the island alone. He said, "The mud was two feet deep. Your boot went straight down to the knee. We could barely get to the cookhouse, and it's only a hundred yards away. We played cards; we couldn't move from the house."

Storms came and went for months. In one lull I caught a ride out to the island with a pilot who was bringing needed supplies. It was clear on the way over, but just after we landed, the rain let loose and the pilot and I ran for cover. Bill said, "By the time summer comes, we'll all have webbed feet." That night I saw lights on in the barn. He said, "That's the vaqueros braiding rawhide, making reatas and reins. Nothing else to do until we can ride."

In May, Nita and I returned. "It's about time you got here," Bill growled, grabbing Nita's and my saddles from the plane. "This was the worst day I've ever had. Trying to gather six hundred head of steers with no help." But when Sam, my dog, wiggled out of the plane and lay on his back on the grass in a submissive greeting, Bill's face softened. "Hello, down there," he said, and bent to rub the dog's belly. The green was explosive. "It puts your damned eyes out," Bill said. When I asked if they'd been lonely, he gave me a dirty look. "Hell no. We make our own fun out here. We don't need nothing else."

In the morning Nita and I were sent up the side of Black Mountain and down Green Canyon to look for steers. "You take the left side and I'll take the right," she said. "We'll meet this afternoon at Arlington—down there." She pointed to the far west end of the island, then rode out of sight.

As I began the long traverse down-canyon, my horse's foot slipped, and she stumbled to her knees. I picked her head up with the reins and kept going. But the ground loosened, and the trail disappeared under a landslide. The rains had been heavy—more than twenty-five inches in a month. I got off and led the horse—first up and around the top of the slide, then back down. The horse fell once and I fell with her, digging my spurs into the dirt for traction. Another time I grabbed her tail for balance as we "tiptoed" across another slide.

Five hours later I arrived at Arlington Camp. Nita loped over. "See any steers?" I hadn't seen a single animal, but I sure got dirty. I was hoping no one had seen that I'd gotten off my horse. It's one of those old rules: you stay on horseback always. But when Jesús rode over, he said, "Is okay. I was watching you. It was *muy peligroso*. What you did was right, getting off the horse. I knew you would be okay." He grabbed my hand and squeezed it; then together, we "caught" the incoming cattle and held herd.

The grass was good and continued to be good. Nothing was overgrazed. Cattle were moved often and the endemic animals—the spotted skunk, jumping mouse, and island fox—our co-residents, lived off the occasional dead animals, such as a steer that fell from a cliff, and all thrived.

We rode long hours in fierce winds. Except for Nita, the rest of us were in our fifties, sixties, and seventies. Each morning

Bill decided who would ride with whom and on which circle. Jesús and I were sent to the Carrington pasture along the cliffs; other days we rode to the lee side above China Camp. Strong camaraderie eased our stiff bodies at the end of each day. Jesús insisted on unsaddling my horse. We passed a bottle of Advil.

The island had always been as self-sufficient as possible. Citrus and vegetables were grown. Supplies were brought when needed by the boat. Water was carried down from Clapp Spring, an ooze that came out from under a rock high up above headquarters on the east side of Water Canyon.

Al Vail, his brother Russ, Bill Wallace, and Jesús made sure the island was a sanctuary for all who worked and lived there. Bill prided himself on occasionally taking in young Mexican kids who were in gang trouble. He gave them a ranching life of long days, teamwork, and laughter, respect for the land and the horses. "So far it's always worked," he said. "We never needed a jail out here. It was the most beautiful place they'd ever seen, and maybe the first time they'd ever felt safe."

Every day on the island, the eye grew wider and the mind. A goofy tenderheartedness surfaced. I found Bill feeding his dog Sissy at the table with a spoon. "Don't you dare tell anyone!" he implored, laughing. Russ Vail raised a raven that had fallen out of the nest. Bill had a pet elk. Meredith claimed one of the steers had fallen in love with her—she could call him from out of the herd and he'd come running.

The daily requirements of tending to land and animals—the basic discipline of a ranch—propelled us through the days; we worked hard and happily and lived in quiet simplicity. In the afternoon I cut the vaqueros' hair or wrote letters for them or cooked. The windbreak of eucalyptus was the nesting place for island ravens. Meadowlarks nested in the grasslands. Horses were bred and born in Wire Field and Old Ranch Pasture and were trained by Jesús and Bill. "Those island horses and me and the vaqueros—we're all the same," Bill said. "We don't know anything but this island."

After 102 years of continuous operation, the Santa Rosa Island Ranch would be forced to discontinue its operation when Channel Islands National Park was created. The Vails and Vickers knew they would be forced to sell and made what they thought was a viable deal with the National Park Service to take over the island in 2011, when their reservation of use and occupancy and special use permit would expire. But the National Park Service abrogated the agreement and, in 1998, ordered all cattle and horses be removed from the island.

It was May of that year when Nita and I flew to the island to help gather and ship cattle. Nita and her family had been litigating a lawsuit against the Park Service; I'd just flown in from Greenland. We were bound by "ranch grief," by the deep malaise caused by an unexpected rupture with the environment one has loved.

The mood on the island was dark. Calendars were being marked off. "How many days?" I asked Bill when we went for coffee before dawn. He said he didn't know and changed the subject: "Some kind of weather is coming. About two weeks before a storm the cattle get restless and move down off the mountains." Then he gave me a withering look: "I don't have the faintest idea what I'll do when this is over. When you've been out here for fifty years, nothing else sounds too good."

Cold. No heat in the house. The cookhouse lights were on and the vaqueros were drinking coffee. The wind speed indicator showed gusts between forty and sixty miles per hour. We walked to the scale house, where cattle would be weighed. The sky was dark until the sun pushed up out of the roiling ocean. There was movement in the corrals: the vaqueros were silhouettes on horseback with their stock whips, pushing cattle into the alleyways. Inside the scale house a bare bulb hung over Bill's head. He readied his calculator and notepad, lit a cigarette.

"Good morning, everybody."

"Buenas."

"Okeydokey. Give me ten steers."

Load after load of cattle were weighed and sent into a corral, then down the wooden pier and onto the *Vaquero*. Then the next ten were weighed and boarded. "We shipped three and a half million pounds of beef off this island last year," Bill said. "That's a lot of food for people. We didn't hurt the grass. We looked after everything. This is one of the best ranches in the world and they're closing us down. Some things don't make any sense to me."

The sky changed. Eucalyptus branches were bent down by bludgeoning gusts. Nita, Arturo, and Pancho took the last of the cattle to the boat, snapping whips as they thundered down the wooden pier. "Once, when one fell off the pier, Jesús jumped in after it and swam it to shore. Loaded it again," Bill told me. "Jesús always takes care of things the right way. He doesn't hold back."

When the last steer ran by, Bill said, "That's it." Jesús winched up the loading chute and the *Vaquero*'s engines rumbled. "It's the beginning of the end," Nita said, wiping tears. The captain waved goodbye as the first of many boatloads of cattle headed for the mainland.

*

EL DÍA LARGO

By the end of May the grass was sun-cured and brown, meadowlarks were nesting, ducks at the estuary of the pasture called Old

Ranch were laying eggs, broodmares were foaling, and half the cattle—about fifteen hundred head—had been shipped off the island.

A ring around the sun indicated bad weather. The wind changed and rain came mixed with sun—a summer shower that misted down all night. We began what would be a month of riding to gather cattle. We climbed canyons, rode the beach, traversed ridges, and the cattle came in brown rivers, down over the hills of dried grass, and streamed into Arlington Canyon, filling the corrals there. Jesús and I closed the gates behind them.

Dust powdered the backs of the animals. We tied neck scarves around our mouths to keep the dirt out of our lungs and leaned against the walls of the line camp in the hot sun. It was June 21, summer solstice. My lips were parched and bleeding. Jesús lit the barbecue and cooked steaks. *"¿Crudo?"* he asked. *"Medio, por favor,"* I said. Nita cut cabbage, jalapeño peppers, onions. She berated the Park Service for the injustice of removing the animals thirteen years before they were meant to be gone. "I just wish they'd waited until my father was dead before they kicked us off. It's killing him. It's killing all of us."

"You can say that again, sister," Bill said. He stood in the open door of the shack and pushed his cap back on his head. "I think it's a good day to give this island back to the Indians. I'd rather not be part of the human race. There are too goddamned many people on this planet."

All morning Pancho had been singing to the cattle as we moved them out of the mountains. Now everyone ate in silence. The vaqueros had no idea where they would go after the ranch closed down. They had come from small villages where there was no electricity, few cars, no running water, no work. José was born on the island; Jesús had plans to return to his small ranch near Tecate.

By evening, the *Vaquero* had returned and anchored at the pier. Jesús and I walked down with plates of dinner for the captain and his friend Sonny Castagnola. "Today is called *el día mas largo,*" Jesús said. "Solstice?" the captain asked. Jesús nodded. "The biggest day of the year."

In the morning we met at the corrals and Bill gave the usual instructions about where we'd ride that day, and who was to ride with whom. He talked about the months to come: "What's coming up is the last roundup. Here's how it will go: first we'll gather Green Canyon, then Pocket Field, and upper Lobo. Then we'll walk the cattle home on August first."

He looked at me: "You can ride with me if you want. The long slow circle. We'll ship on August second and third, give the captain a few days off, ship again on the seventh, eighth, and ninth. Then it'll all be over, and I'll hang up my chaps and spurs for good."

In July, Nita left for work and I stayed on the island alone at headquarters. The white two-story clapboard house was built in the late 1800s. The floors were linoleum and smelled of mice, and the single-pane windows rattled in the wind. On a day off I went for a hike in Lobo Canyon. The island had been likened to the Galápagos with its native trees, plants, and animals: the Torrey pines, spotted skunks, island foxes, and wildflowers. In one amphitheater, a rock hung from a cave roof like a beaver's orange fang. The year-round stream trickled. Reeds and grasses thickened on the canyon floor, still green and full of island endemics: *Dudleya*, saline *glauca,* and seaside daisy.

At the mouth of the canyon, the creek let out in a long S curve and the canyon floor was littered with bits of moss and seaweed. Tufts of fog flew past and dropped down into seawater like flowers. That night Jesús walked in the dark to headquarters, following the creek behind the buildings so he wouldn't be seen, and knocked on the door. I ran down the stairs. He was wearing a fresh white shirt and stood in the entry with his hat over his heart, and in Spanish, asked me to marry him.

*

EL ÚLTIMO DÍA

I had been gone from the island for a few weeks after shipping the last of the cattle and returned when it was time to gather the horses, all of whom had been born and raised on the island. Flying over the ocean with my dogs, saddle, and notebooks, I saw the lighthouse light at Point Conception turn and sweep the ocean not to see, but to be seen by anyone passing, and because Conception was also the Western Gate, the light showed the way ahead for the dead.

That day of passage to Santa Rosa Island—the last I would make—sea drizzle seeped and pooled into unearthly shapes, and the ocean at high tide ate the edges of the island with ferocity. It was only a month earlier that José had unloaded sacks of beans and rice and a bucket of Mexican limes. He'd held one up: "Each of these are green suns," he said, smiling. "Good for tacos and tequila." By the end of roundup, all the food he had accumulated and cooked would be gone.

By the time Nita and I returned, the moment of vernalization had passed and green pastures had turned bronze. The French doors at headquarters opened to an untended yard, where we sat in the warm sun. Nita's father, Al, joined us for iced tea spiked with I'm not sure what. He talked about the past: "I grew up cowboying. Can't remember when I started. It was always just something I did. Spent summers cowboying on different ranches in California, then went to college at UC Davis and UCLA. I wasn't too happy with college and they weren't too happy with me. I fooled around a lot, drove a car down the steps of Royce Hall, but didn't get caught. Hell, they'd have thrown the key away if they had. I wanted to ranch, not go to school, so I came on the island full time as a cowboy.

"When the war came everything changed. I was 4-F and was turned down twice for the Navy, the Air Corps, and the Army because of my asthma. So I was on Santa Rosa Island all those years. I stayed in the bunkhouse in the end room. An island is a lonesome place, and if you can't handle the isolation, you're in trouble. I like it best out here. I didn't go to the mainland very often.

"There wasn't anything to do out here if you weren't working, but there was always something. In the summer I'd go to see Phil Orr, the paleontologist. He had a lot of fox pets living with him in the cave where he had found an early human site. Fourteen thousand years ago this island had people living on it. Now they're kicking us off. Sometimes Phil would drive his old jeep over Black Mountain to Johnson's Lee, the camp the Air Force kept during the cold war, and go to the officers' lounge there. Then he'd drive all the way home. DUI all the way. Went home by the braille system because that's a damn tough road even if you're sober.

"One winter a ship called the *Chickasaw* wrecked near China Camp and Diego Cuevas and I went over to see her. She'd been running in bad weather and came up on the beach. She was chock-full of children's toys from Japan and we were a couple of grown men sitting on the beach playing with them. All times were good over here. We knew the families on the other two islands, but like I said, island life is hard on people. Carey Stanton came out of Stanford medical school as a doctor but quit practicing. He inherited Santa Cruz Island from his family and couldn't stay away. It finally killed him. Luke Humphrey got killed when a horse fell on him. He'd been chasing a fox for fun. Stupid thing. Died on the boat on the way to town. We had ghosts over here too. Mrs. Pepper walked up and down the stairs, but only when the wind blew. It was the wind that drove her crazy. Over on San Miguel, Herbie Lester ran sheep and built a fabulous damned house out of driftwood. But he was shell-shocked in World War I, and in the end, he killed himself.

"Jesús Bracamontes and I were good friends always. I can't remember when it started. It just was. I called him Brac. He pulled me out from under a horse in the Lepe country and I pulled him out from under a few too. He'd retired from cowboying because he had a bad back. Had it operated on and the doc told him he could never ride again. But a few years later he called asking for his old riding job back. What the hell. 'Sure,' I said. See, I'd helped him out with the funeral expenses for his wife, Virginia, when she died, and he wanted to repay me by coming back to work. He didn't have to do that, but that's how he is. I guess you could say he's riding the island for Virginia."

Later that day Nita and I drove with Bill Wallace from pasture to pasture looking at the horses. "There are a hundred fifty head of horses on the island," he said. Mares, foals, yearlings, and the older horses the cowboys use. He stopped to show us the two-year-olds he'd been training. "You're not going to get a horse to like you until you start thinking like a horse," he said. "They'll know when that happens. People are always trying to put them in high school before they've been through kindergarten. They're like kids. You have to take them slow. We lived out here like a cat with nine lives. We made our own fun. Rode colts on the beach. We didn't have anything, just the animals and ourselves."

We drove over a hill down into another pasture. Above us was the stand of endemic Torrey pines. An island fox trotted by, looked at us, and kept going. We bounced back down to the corrals, but on the way we stopped to look at the pasture with the old, retired horses. "They'll go to the canners. They don't deserve this. They should be allowed to retire here after all they've done for us. These horses have given their whole lives. But the park doesn't want them here. No one wants them." He put the truck in gear. "Goddamn, I just can't look anymore."

There were horses in corrals and more coming down the hill. Bill had refused to help gather them. "The thought of getting rid of these horses is eatin' a hole in me."

The next morning, we were up at four. "This is *el último día,* isn't it?" Jesús said. The last day. Before dawn, Jesús and Pancho had saddled up and begun bringing the horses in. We watched them gallop down the steep slope behind the cookhouse, cross the creek, then funnel into the corrals. Bill hadn't gone with them. Said he couldn't do it. Wouldn't do it. Wouldn't have anything to do with taking these horses off the island. "They were born here. That's all they know."

By the end of the next day the cutbacks, the pensioners, the mares with colts at their sides, the snaffle-bit horses, the saddle horses mingled in the corrals. Bill couldn't stay away. "There's Muñeca, Baby Doll, Lola, Roberta, Machichi, Flicka, Chuey, Mariposa. . . ." He walked through the corrals naming each horse, remembering something special about them.

The vaqueros caught six at a time out of the herd and led them under the eave of the shed. Meredith filled syringes while the cowboys curried the horses tenderly and combed their manes and tails. Finally, Bill joined in, sweet-talking them as he gave them shots. "Hello, Major, hello. Goddamn, that's a good horse." He turned to me: "I started him but had too many horses to ride so I gave him to Arturo, and he turned out good. There we go, that's a boy. You're a hell of a boy. Okay, next.

"Now we've got Samba. She's a little crazy but kind of funny too—whoa, sweetiepie, whoa. Are you going to stand up here? That's a girl."

The breeze was cool. Bill gave two shots to each horse. Fog wafted out into the channel, leaving the red barn in sun. Pancho doused a cigarette with spit and led a horse to Bill.

"Here's Chili. How's old Beaner-boy? Beanie, Beanie, Beanie—whoa . . . He kind of acts silly and skittish but as soon as you get your hands on him, he just dies down real quiet. I'll miss him. He was born in '83 and he still won't let me give him a shot. Whoa, sonny boy."

Jesús came into the shed with a mare. "Hello, old girl Ritzy. We made a lot of hard miles around that south side, didn't we? Now you're going to town. . . . What's the matter, girl?"

And another mare they were still trying to catch. "Look, no one can catch her!" Bill said, laughing. "Goddamn. Okay, here we go," he said when Jesús led her in. "How are you?" he asked the horse. "We've done a lot of things together, haven't we? You're not going to like this. You're a good girl. Everything's all screwed up here—you're going to town and you'll buck someone off there. Wow, can she buck! Yeah, you'll give it to them good."

Ravens crossed over the barn, flapping hard as if trying to erase what was happening. The last horse was brought to Bill. He pushed his hat back and scratched his forehead, then took the lead rope and stroked the horse's muzzle. "Hi, old Marty. Well, we've seen a lot of cattle, haven't we? We've done a whole bunch of everything. Now, no more China Camp, no more Pocket Field, no more island. . . ."

In the late afternoon the horses were turned out to graze and fog stood in the channel like a wailing wall. The horses were the third generation of the original horses Walter Vail brought to the island from Arizona more than a hundred years ago. The family of animals and humans was being torn asunder, and a century's worth of hard work, love, and nurturing was being lost.

I couldn't sleep. Neither could Nita. All night a sorrel gelding galloped back and forth along the low fence near the ranch house, whinnying for the mare he loved. The windbreak heaved and sighed. Sometime after midnight Nita and I went to Bill's house. He was sitting alone at the kitchen table stirring a cup of cold coffee. He looked up but said nothing. We sat with him.

Finally he spoke: "I'll tell you what I'm going to do," he began. "I'm going to take the best old horses over the mountain to China Camp and I'm going to let them run on the beach and then feed them a big pile of grain, and then I'm going to shoot them. And after, I'm going to shoot myself."

There was a stunned silence. No one moved. Then Nita shoved her chair back and stood. "No, you're not!" she screamed, and stormed out the door. That night we drank tequila, walked out to the barn, came back, drank more. In the morning she looked for Bill and found him alive. The *Vaquero* was at the pier bobbing back and forth as the tide came in and the fog cleared. Bill and Jesús came for me in the pickup. As we passed the cookhouse, we saw Pancho alone at the counter sleeping with his head in his arms. "He better wake up," Bill growled. Under the yard light he chewed the end of an unlit cigarette almost in half. His face was hard.

From the corrals Pancho took five horses down the pier. The tide was right, and the loading chute was lowered onto the boat. "Get the hammer and nail up that gate so it won't fall," Bill yelled to Sonny over the roaring engines. "Goddamn, these are horses, not cattle, and we have to make it safe for them."

One by one Pancho led them down the chute. They were wild-eyed and skittish. They had never been off the island. One turned and pushed against the others. "That's what I was afraid of," Bill said. Jesús took a lead rope and started over again. Slowly the horses filled the bottom deck of the *Vaquero*. A plane circled and landed, and I watched as Nita and her father flew away.

"Okay, we're loaded," Bill announced.

The gate on the boat's rail was closed and the lines were untied as Pancho, Jesús, and I jumped aboard. A wind came up as the engines revved and slowly the boat pulled away from the pier. I looked back: Bill stood frozen in place. The sun lifted. I raised my hand to wave goodbye, but he had already turned away.

Out in the channel, waves broke over the bow. Jesús and I sat with our backs against the bridge. He put his hand on mine and spoke calmly to the horses, trying to console them, trying to console us. The breeze was wonderful. Sunfish jumped and sailed like arrows through the air. We passed the knotted west end of Santa Cruz Island and waves splashed against its sun-softened cliffs. I took off my cowboy boots and socks. I had an infected cut in the shape of a cross on the top of my foot, and when Sonny passed by and saw it, he said, "It looks like stigmata," then dipped a bucket with a long rope into the ocean and poured salt water over the wound. "That'll heal it," he said.

Born in Genoa, Italy, Sonny told me that his two maiden aunts had hidden Jews during World War II. Acts of kindness ran through his family. He always brought food aboard—figs and avocados from his orchard—and that day he'd made his famous cioppino. I said I wished we could find a sanctuary for these horses, find someone who would give them shelter.

The horses nipped at each other. They weren't used to being crowded together, and they'd never been on a boat. Pancho and Jesús talked to them sternly and lovingly until they quieted. Then Jesús turned to me: "*Mamita, mi corazón está muy malo. . . .* Come with me to my rancho in Tecate. I will take care of you."

The night before he had found me alone in the barn sobbing, not for myself but for the horses, the vaqueros, for the island itself. "It's getting smaller, isn't it," I'd said to Nita. "All the places we can go to find solace."

Just then the boat was taken sideways by a swell, and for a moment it was like standing on water. Jesús and Pancho called to the horses to calm them as they scrambled for footing. The boat righted itself. Another fish flew by. "Let's go to Mexico," I said. "With these horses!" Pancho and Jesús lit up. "*Sí, sí,*" they said in unison. I looked at Jesús's handsome face—the Aztec profile, strong jaw, and beautiful hands. He was a good man. Nita had played the Lyle Lovett song that goes "If I had a pony I'd ride him on my boat . . ." The idea of running away with this boatload of horses to live in Mexico began to seem less and less farfetched.

The sea rose up under us with dappled spark-light and leaping dolphins as we headed for Port Hueneme. This was just the first of several boatloads of horses we would deliver. The name Hueneme is derived from the Ventureño Chumash phrase, *wene me,* meaning "resting place." A euphemism perhaps, as we knew that many of these horses were going to die.

On the other side of the Potato Patch, where the sea was rough, the waves evened out and we went downhill, pushed from behind by swells whose slow monotony took us like Charon's boat all the way to the other side of the channel.

Part III

II.

A Buddhist monk who carried a walking stick made from a tree branch said to me: "Your life is like this rootless tree." Back from Santa Rosa Island, I had trouble feeling comfortable in my own house. I longed for the island, the vaqueros, the communal meals, and the horseback loops from ocean to mountain. I walked the ridgetops of Hollister Ranch looking south toward the island, but a fringe of clouds kept obscuring it from view. On restless days the floor of my house seemed to pull out from under me and eventually, the friction caused me to flee. The match lit by a lightning strike years earlier still burned. The monk quipped, "It must have a long fuse."

*

It was the beginning of May and the start of the dry winter season in Africa when I arrived in Johannesburg. I was to meet Allan Savory and travel with him to Dimbangombe, the game ranch in western Zimbabwe where the Africa Centre for Holistic Management is located. It was an unlikely oasis of calm, self-sufficiency, and plenty in a country on the brink of famine, death from disease, drought, and political chaos because of an out-of-control dictator, and I knew that Allan was restoring land and lives on his native ground and working to stop the desertification of the planet. Outside the plane window, the sky was black, but under a full moon I could see the sandy northern part of the continent give way to the gnarled overgrowth of the Congo, then the deteriorating grasslands and bushveld of central and southern Africa. Recently Zimbabwe had been pronounced "hopeless" by a writer at *The Economist.* I wanted to see for myself and walk the ranch that stood out as an island of sanity and safety in anarchic times. At first light I bolted from my seat and knelt by the tiny round window of the plane's aft door. The knot of Kilimanjaro was shrouded in storm clouds. I was flying down the first human's long spine. When Pangea's crust split, Africa was pushed southward, escaping the ice sheets that covered Europe and North America, and it was easy to see why the savannas of eastern and southern Africa were the birthplace of human beings.

Allan met me at the airport hotel. Over dinner he warned me about outbreaks of violence. Residents of Joburg had installed "rape gates" in their bedrooms after a rash of horrific break-ins, and things were even more tense in Zimbabwe. Robert Mugabe had been lashing out at what he called "barbaric British colonialism" despite his association with Allan after the civil war and the fall of Ian Smith's racist regime. Mugabe was redistributing land, putting "war vets" out in the bush. "That war must be finished," he kept saying. Allan had fought against Ian Smith as well. "Mugabe knows that as a member of Parliament, I crossed the floor to oppose Smith, and he knows I share my land and resources with the villagers. He would never harm me," Allan said. I hoped it was true. But Mugabe had already targeted three thousand of the five thousand white-owned farms for confiscation without compensation and had put heavily armed ex-ZANU-PF soldiers on land near the Africa Centre in what Mugabe called "a justified protest against unfair land ownership by whites."

On the early-morning flight to Victoria Falls I looked down on ancient sand dunes, bare-ground villages, water holes where elephants drank, dry pans, and wandering cattle, humans, giraffes, Cape buffalo, kudu, and impala. Mud huts with thatched roofs were gathered in loose villages. Game trails splayed out through the bush. Animals and people were walking. Born in what was then Rhodesia, Allan was the son of a civil engineer and grew up in the bush. "As a youngster, my only aim in life was to live in the wildest African bush forever. I did have that for a time, but I ended up forsaking it to work toward saving the wildlife that had drawn me to the bush in the first place." Over the years he enlarged that context and became a restoration ecologist. He now works globally toward the regeneration of vast grasslands in order to bring life back to degraded soil and reverse climate change.

Civil war erupted and Allan's work of saving wildlife in Rhodesia was no longer possible. Later, he was elected to parliament and took the lonely stand of opposing Smith. For that he was subjected to nine assassination attempts. "Finally Smith's thugs threw a petrol bomb into my house at Kazuma. I wasn't there, but they thought I was. I had a lot of artillery under my bed and bullets started flying. My house was made of mopane logs set upright and chinked with red mud. Some inside walls were mudded smooth and whitewashed to give more light. The roof was thatch. I'd cut holes in the walls for a door and windows but had no panes—they were open. Nothing was closed: animals could come in and out. They tried to kill me but found out it was going to be a much harder job than they thought—I'm still alive!"

Allan was forced to make a dramatic middle-of-the-night escape with no passport and no money. "I went to the airport, filed a flight plan to South Africa, said goodbye to my children, then flew under the radar to Botswana. There, I was able to hop on a British Airways flight. They let me get on with no ticket." He ended up on a flight to the Cayman Islands and had no choice but to live there until things settled down. He speared fish and lived in a shack on the sand. It was there that he had his dark night of the soul. "I lay on the floor for a long time wondering what to do." His marriage had failed; his lover visited but quickly left, refusing to live on the lam. He had no money, no country. But he had worked as a restoration ecologist and put his shingle out as a consultant. A wealthy Mexican rancher called, and his new life began. Other ranchers welcomed him, and his consultancy moved into Texas. It was there he met Jody Butterfield, who had come to interview him for a magazine, and not long after, they married. "It's been a damned long interview," Allan said.

After the civil war ended, Rhodesia became Zimbabwe and Allan was invited to Geneva to work with Robert Mugabe on the peace accord. In those early years of his regime, Mugabe welcomed the presence of nonracist white farmers and game ranchers like Allan, who had become a hero to the black African soldier, but that initial inspiration soon faded, and it was a long time before Allan could go home. Looking down from that early-morning plane, Allan said, "I can really only be happy here." He pointed at an open stretch of savanna and thornbush. "Look, there's Kazuma, my old ranch," he said. "That's where I spent the best years of my life with my family. I wasn't home much, but when I was . . ." He stopped and gulped. "There's nothing better than raising children in the bush." He peered down excitedly: "We had nothing, no electricity or phone, and we were barefoot all the time. The game and the children wandered in and out. One night a lion slept on the ground next to me and I never even woke up! We were as wild as the animals."

Now that forty-eight-thousand-acre ranch is part of a national park. "It was taken after the war and all I had left was a smaller game ranch that I quickly put into the hands of African people. I had realized that if Africa was to survive at all, it had to become part of a functioning multiracial society. Together we formed the Africa Centre for Holistic Management, a place to learn how to make decisions about the natural world and society, about managing holistically. I hired Huggins Matanga to manage the Centre. Recently retired, he has been there since the beginning, a natural statesman respected by the people in the communal lands. Our mission is to share land and resources with villagers across the road and to restore the grasslands and make the land and the society healthy again." As the plane descended, Allan pointed to another place on the ground. When I leaned over to ask what it was, he averted his head. He was wiping away tears. "My son Jim died in an accident. That's where he's buried." The plane tipped and we circled: something white billowed. Then I saw a mile-wide gash: Victoria Falls.

We were driven to Dimbangombe from Victoria Falls. The side of the road was crowded with people. Women carried water jars on their heads, teenage girls walked hand in hand, mothers with babies walked single file toward town, and a cart, pulled by donkeys four abreast, was stacked high with beer barrels. Baboons cavorted in towering acacias that lined the highway. There were few cars. Open-air roadside stands displayed crudely carved giraffes, but there were no tourists to buy them. Because of the political unrest and the killings, outsiders were afraid to come to Zimbabwe.

A two-lane highway divided the Hwange Communal Lands from the Africa Centre. One hundred and forty-five thousand traditional people lived on a million acres of badly deteriorating, bare land. As soon as we turned off onto the road dappled with elephant tracks as big as platters, the vegetation became lush. More tracks showed that giraffe, kudu, sable, impala, and dik-diks had been there. The ruts were impregnated with coral-colored sand that had blown in from the Kalahari. On either side was a thick forest of mahogany, msasa, weeping wattle, sickle-bush mopane, and umbrella thorn. Five long-beaked hornbills flew up and an elephant, barely visible through a screen of tall thatching grass, lifted its trunk and delicately pulled fruit off a marula tree.

Here and there the view opened out into small valleys of tall grass just beginning to go brown. Ahead we could see a herd of African cattle grazing with three herders alongside. The name Dimbangombe means "the place where cattle get lost in tall grass." What too few understood then was that overgrazing has to do with time, not numbers of animals. If cattle are moved every few days, there is no overgrazing, and even in a drought, the grasses keep growing.

The sun was hot, but soon it would be getting cold. At latitude 17° south the equatorial winter of bitter nights and days of hard sun had just begun. The Africa Centre was perched on a hill above millions of acres of wild game land, woodland savanna, bushveld, and thorn scrub interlaced with grassy valleys. A huge rondavel sat centrally on a lawn. Thatch-roofed and circular, with a low wall and a firepit in the center, it was the classroom, dining room, and community center. Beyond were rows of raised boxes where vegetables were grown.

Matanga greeted us. Small and rotund, he carried himself with a sophisticated reserve and wasted no words as he briefed Allan on the latest political situation: "The war vets have pegged our land. We had a very good meeting with them and with the elders. We will try to help them." They both knew that if degraded land was restored, there would be enough food and game for everyone.

After a late lunch we were driven down to Allan's camp on

the banks of a small river under a big tree. Allan's tent was a large fly hung over a bamboo ridgepole. He placed his bedroll in the center and unrolled mosquito netting. To one side was a stack of metal trunks, and on the other side was a gun safe with four rifles. When I threw my bag onto the cot in my tent—it was big enough for two small beds, a table, and a lamp, with screen windows—baboons scrambled, and a curious vervet monkey peered through the window at me.

Small-boned and wiry at age seventy, Allan quickly changed into shorts and a shirt and took off his shoes. He grew up barefoot and wasn't about to change. Sent to boarding school at the age of seven, he said, "The first thing I did was ask an African family nearby to keep my guns. In exchange I went hunting every morning before my first class and shared the meat with them."

Allan has an Englishman's dry wit and loves nothing better than a practical joke but is easily brought to tears by injustice to an animal. He has adopted baby elephants and warthogs and brought abandoned horses to Dimbangombe as white- and black-owned farms were confiscated by Mugabe's troops. "To be part of a hunting and gathering society is to learn how to be a human being," Allan said quietly. He knew that sick land meant a sick society, that the loss of biodiversity meant the end of life—and he fervently wanted Africa to survive.

The previous week a villager from across the highway walked his six cows from the village to join the Centre's herd. He was nervous because the cattle were his only wealth, but it was the Centre's policy to replace any animal that died with one of its own.

We sat around a smoldering fire. "Big Boy" Nyatu arrived with a fresh load of firewood and stirred the embers into flame. Allan greeted him speaking Ndebele, and the two men squatted, pulled a grate over the fire, and boiled water for tea.

"Sharing land and resources with the villagers means convincing them to bring their livestock to join the common herd. The ground in the villages is completely bare and hard-packed from so many animals running loose—cattle, donkeys, pigs, chickens, plus wildlife. Their boreholes are going dry; they've run out of thatch grass. We want to help restore their communal land so they don't starve. Fifty percent of villagers are dependent on UN food. Some have small gardens, but the elephants come in and trample everything. Bringing life back to the soil, reinvigorating the water cycle, restoring the grass, and planting gardens will result in prosperity: they will have water in their wells, cattle for meat, vegetables and fruit, and a healthy society. You see, poor land leads to poor people, conflict, disease. To heal the planet, you have to heal the whole."

After settling in, I followed Allan during a walk. The land was thorn trees, rocky knobs, streams, and open vlei. Allan carried a rifle and trotted barefoot through the bush, his toes bending gracefully over stones and thorns, pushing into the red dirt. His brown eyes were by turns keen and mischievous. We crossed the creek and walked out through waist-high grass. It was the only place for miles around where the soil was healthy. The cattle were nearby with their herders. They'd been grazing there for a few days, and would be moved to a new area in the morning.

As a game ranger in northern Rhodesia, Allan observed how large herds of wild buffalo and antelope, tracked by prides of lions, kept the grazers tightly concentrated. The grasses depended on the animals for dung and for trampling in the seed. Their defense strategy—to eat and move quickly in enormous herds—was the key to survival. That's what planned grazing with livestock replicates today.

Allan and Matanga made a point to hire as many local people at the Centre as possible, and the herders were deemed most important. "We give them guns, clothing, food, and even shoes if they want them," Allan said, smiling and wiggling his bare toes. That evening we walked to find the cattle. The grass was thigh high. We watched the animals being herded into the kraal—no whooping or yelling—as the cows pressed into the enclosure. There were nine herders, including one woman—and two black-and-white smooth-coat fox terriers to protect them. The kraal was heavy but portable. Made of sticks and bound together with vines and rope, it could be rolled up and moved.

At night the herders lit small fires and slept on cots in tents around the animals. Any disturbance woke them—a lion's roar, or unusual jostling of the animals, or the dogs' barks—and the herders would make sure no leopard, wild dog, or lion got into the livestock. In the morning the animals were moved out to new areas to graze according to the land plan, and the kraal was moved. Planned grazing told the story. In the middle of millions of acres of bare land, Dimbangombe, true to its name, had tall grass all year.

On the way back to camp I stepped in Allan's barefoot tracks. Thornbush snagged our clothes. Following a fresh trail, we climbed a mound of red rock that caught the evening light. An albizia tree was full of nutritious seedpods, but another one had been pushed over, its roots exposed. "Elephant," Allan said. On the track we saw white hyena scat and the curved tracks of an aardvark, an animal so shy Allan had seen only a few of them.

A group of impala barked. At the river a malachite kingfisher with blazing blue wings fluttered and dived. It was difficult to remember that not far from here the ground was bare, streams were waterless, and people were hungry. We crossed the creek barefoot and an oxpecker rose up squawking. The water was still warm from the afternoon heat, but as night came on, the air turned deeply cold.

A chocolate hawk circled. The sun was going fast when we heard snorts and movement. "Cape buffalo," Allan whispered. "If he's in a bad mood he can be awfully mean. It's the only animal I know that will come back and ambush you." We walked carefully, quietly, and continued on. In a clearing a kudu bull stood near his harem of six females and their calves. His spiraled horns looked like driftwood—sea-smoothed and curled under by riptides.

Near camp I took off my boots and walked barefoot. Allan laughed when I winced at the thorns. Acacias swayed in the evening breeze; baboons called and squabbled. It was only six in the evening but almost dark. Perhaps the "darkness" Joseph Conrad alluded to was no more than this rapid-fire fading of day.

*

In the morning, bundled against cold and mosquitoes, Allan talked about the civil war. "When Ian Smith imposed censorship on the press during the war, *The Herald* came out with blank pages in protest. Smith said they were fighting for survival, but they weren't. We were fighting for independence. He said he was the last bastion of civilization in Africa, but he was doing things we'd fought in World War II to stop. To oppose Smith was illegal. I joined Parliament, and when I declared my opposition to Smith, a billboard was put up in Harare: TRY SAVORY FOR TREASON.

"After the war my ranch at Kazuma was expropriated. It was three hundred miles square. A kangaroo court set up by Smith established the value of my land. I had a lease option on it, but I was not allowed to exercise the option for six months. They said that with my war record, I would be dead in six months anyway. They left me a small portion of that land and when I tried to make it a research station, they refused, then they just took it from me. After, they seized all the ranches that belonged to nonsupporters of Smith. Now Mugabe has turned land redistribution into political roulette.

"The exodus of successful black professionals and farmers is greater than that of the whites, and we can't afford to lose them. Many of the white farmers here were never racists. They could have stayed and helped support the infrastructure for everyone. But as more black professionals leave, the day-to-day government gets weaker. Good government is not possible until there's a sound land policy. You can't go around looking for a perfect democracy with hungry people and land so degraded that nothing grows." At night we listened to the BBC World Service on Allan's shortwave radio. There was news that the head of a Canadian NGO had pulled out after he was attacked, and as a result, five thousand war orphans were left out on the streets with no food or clothing. "Now all the NGOs are leaving," Allan said. "Corruption is so rampant, no one wants to come here. That's why we employ only locals here—lots of them. So much of aid is damaging. It's always top-down management and never sustainable. We returned a grant from USAID because they wanted us to hire Americans. Why would we do that when there's ninety-eight percent unemployment here?"

We put more wood on the fire, and the resident gray parrot peeked out from its hole. A large family of baboons scattered the monkeys who had come to beg for food. "Over there, under that big tree, is where a lioness likes to sit and watch us. Keep your eye on that spot and maybe we'll see her," Allan said. Upstream, two impalas drank and the fish that give birth through their mouths wriggled away. In the morning we joined the others in their final classes in Holistic Management. They had come from all over South Africa. There was Douglas, a traditional Herero from Namibia; Moses, an ethnobotanist–traditional healer; Colin, a white farmer from Namibia; Christine Jost, a veterinarian from Tufts University; plus two teachers from the Centre: one from South Africa and the other, Elias Ncube, from a nearby village.

On the rafters of the big rondavel, two small owls copulated over our heads while Allan talked. He sneaked a look up at them from time to time, smiling. Red, a horse belonging to Matanga, stood inside for the shade.

"Life is complicated; Holistic Management is simple. It is all just common sense," Allan began. "It's much less about managing than about decision-making that results in seeing the whole and restoring and enhancing the natural world."

There were discussions concerning biological health and quality of life, the basic ecosystem blocks that made photosynthesis possible, and testing guidelines that guided decision-making. "You have to get down on your hands and knees and look!" Allan said emphatically. Later he told me that holistic management is easier to teach in places where people don't use money as armor to avoid learning new thinking.

After dinner we gathered around a fire to tell stories, but when a jeep roared up the long dirt road, everything stopped. "It's no one we know," Matanga said in a low voice, and he told Richard and Albert, two of the Africa Centre workers, to greet the visitors. Five shoeless young men in tattered clothes jumped out. They were carrying machine guns.

Allan stayed seated. "It's best if I let Matanga, Albert, and Richard deal with them," he whispered. "The old war vets know me because I fought with them, but these kids aren't old enough to have been in the war," he said, glancing again at the men.

They were talking animatedly. "We give them a gentle reception always. I don't resent them at all."

The men looked around, guns still slung over their shoulders, then climbed into the jeep and left. Matanga came back to the rondavel. "They want to know if we are helping the villages," he said. "And I told them what we are doing here. They said they will come back tomorrow to see."

Allan warmed his toes by the fire. "These settlers have been taken out of the city and put on wild land by Mugabe. They have a lot going against them. The land is full of dangerous animals and snakes, and the ones who were resettled during the rainy season had no shelter or gardens. They made do and in the long run were glad to be here. Their needs weren't taken into consideration by the politicians, but it might work out anyway. For their sake, I hope it does."

After a dinner of locally grown salad, sadza, and roasted meat, beer was passed along with Moses's special herb tea. "I'm from the University of Life," Moses declared with a deep laugh. "We can cure the incurable!"

A mother pig and three piglets wandered through the rondavel and were shooed out. There was teasing about bride prices in a Herero village, healing potions for failed love affairs, and more stories about animals.

"When I was in Northern Rhodesia," Allan began, "the game scouts and I lived out with the herds. I had a gun and a sack of tea and one of sugar. I used a dead elephant's ear for shade and shelter. I've run alongside rhinos. They have such a graceful run—their head stays still. It's not too dangerous because their eyesight is bad. Sadly, they're mostly gone now.

"One night I was sleeping near the open doorway of a friend's house where his dog was lying. A leopard leapt in and snatched the dog and that was that. Fast and quiet. She'd been watching us for a long time, waiting for me to close my eyes. Another time a lion caught a cow, jumped over a brush fence, and ate it only ten feet from where I was sleeping. I never woke up.

"When I was young and living out in the bush, a Cape buffalo charged me, and I leapt on its back. I leaned forward and held her horns, but she bucked me off, so I tried to wrestle her from the ground. I was losing. Just when I thought I was done for, my partner roped her back legs and yelled, 'Let go!' and pulled the animal out from under me.

"That happened when I was in charge of Northern Rhodesia's game, and I did a lot of tracking. When you're tracking you have to think ahead and see where the animals might be and where they'll go; otherwise you'll get stuck. Your head will be down staring at your feet, and you will have missed the whole picture. Tracking means running. I could run down a zebra—they run slowly enough to keep you in sight. They'll look back, and when they break, you're close enough to run at an angle straight into them. Hunting eland, I ran alongside them and shot. We often ran for twenty miles or so, up and down these hills, through the thorn. You had to be fit," he said, smiling.

Later that night the talk turned to AIDS. Twenty-five million people had died of the virus in Africa before the antiretroviral therapy arrived. At the Africa Centre some of the staff was getting sick. Allan explained: "It's against the law for a doctor to tell a patient that he or she is sick. People with AIDS are shunned, put into a dark hut by themselves almost untended to."

Elias Ncube, a well-educated local villager on the staff, speaks five of the local languages. He is rail thin and philosophical, kind and gentle. "The government has imposed censorship on every outlet, so people aren't educated about how they are getting AIDS. Inflation is out of control. Even if there was medicine, no one would have money to buy it. There is no fuel, no food, no commerce, and therefore no employment. Many people, young and old, are dying of AIDS. We go from funeral to funeral. There are no weddings."

Late in the evening tea and biscuits were served. Allan lost his first marriage during the war, and three of his children have died. "But marrying Jody was the best thing that happened to me." He looked around, smiling. "I can talk about her when she's not here. What I love about Jody is her innocence. She doesn't have a deceitful bone in her body. She's been my helpmate, and without her, none of this would have happened.

"And Matanga . . . well, he runs the Centre. He understands Holistic Management perfectly and helps with the deeper cultural parts of African life that I would have missed. Earlier Matanga said he'd have to make a run to Botswana the next day to get fuel. It's against the law to bring gasoline in from neighboring countries, but Matanga has been sneaking it in through the checkpoint at the border, and so far, it's worked. Zim money is worthless, there's no food in the stores, and no fuel. That's how a dictator keeps the people under control," Allan said.

The fire burned low and no one added more wood. It was getting late. Back at camp Allan lit a kerosene lamp, set it on the table. "In the old days there was more of everything—more animals, trees, grasses, bugs," he said. "During the February rains there were hundreds of fireflies, thousands of moths. We counted 132 species of native grasses, over 21 species of antelope, and 50 birds that are endemic to this region. Now there is less of everything." Lying on my cot, I wondered how Allan kept going despite the undoing of a country he so desperately loved. Few had lived as he had. Few had the guts to encompass the global breadth of his thinking, then put it into action. He dared to tell people that saving single species was a waste of time—you had to make the land healthy before anything and anyone would survive. He dared to say that UN aid to villagers kept them from planting crops. He dared to say that the root cause of climate change is the desertification of the planet, degraded land, and ineffectual rainfall. He dared to say that planned grazing could save the planet from the climate catastrophe, that if only half the people of the world would manage land holistically, putting carbon back in the ground, we could take things back to pre–Industrial Age conditions. He dared to say there may not be enough time.

*

A swayback female dog so thin she seemed to be only a skeleton and a row of teats skulked into the Centre's new permaculture garden, and was shooed away by little boys throwing stones. "It's been a long time since I've seen a dog that thin," one of the staff said. "Yes, they are getting hungry and the children in the villages are too," she added as she poured water from a bucket into a watering can. We followed her between rows of broccoli, kale, carrots. Dressed in a red skirt, pink blouse, and yellow headscarf, she was barefoot and mud squeezed between her toes. "We are now at the beginning of a famine and we must grow as much food as we can. And we will!" she said enthusiastically. There were round beds and S-shaped beds. Every plant was hand-watered.

Elsewhere farmers were burning fields to get the "green flush." Ash provides a natural fertilizer, but Allan was discouraged. "Fire pollutes the air and creates even more bare ground. One hectare of land burned equals the emissions from more than six thousand cars, but they continue to burn more than a billion hectares in Africa per year." He turned to me: "You see, it's right here that we are causing your beloved Greenland ice to melt!"

*

In the middle of the night a truck careened down the road and stopped at the Centre's rondavel. Allan was summoned quickly, and I went with him. Billy, a young African farmer, stood disheveled and exhausted. He had been to one of Allan's trainings and knew that if he was in trouble, Allan would protect him.

"They came to my parents' farm," he said, breathlessly. "They came and pulled us out and put us in a shed; then they put all the cattle inside with us. My parents were killed. I got out and had to run and hide. I've been traveling all night. I came here to be safe."

Allan put a hand on the young man's shoulder. He was agitated, anxious, grieving, and we stood closely around him. "It's good you came," Allan said.

"They are grabbing three or four farms a night, then reselling them for much more money. We're losing all the good farms. Some black-owned like ours, some white-owned. Black, white—it doesn't matter. They are taking everything that's worth something."

The kitchen help brought a plate of food and tea, and Billy was told he would have a warm, safe place to sleep. "You can stay here, of course, for as long as you wish" Allan said.

Billy continued: "What has been happening to white farmers, now it is the same for black commercial farmers. We are all looking for safer places to go, to live. We're being discriminated against too. Mugabe is stealing from anyone who is doing well."

Back at camp, Allan and I couldn't sleep. Things were coming apart in Zimbabwe, South Africa, and Zambia. Violence was increasing. There were four or five farm deaths every day. Land was being claimed, animals were being poached, trees were being cut down for firewood, and inflation was roaring. At breakfast, Matanga said, "When the Zim dollar and rand are worthless, how will people buy food?" Roving bands of thugs had begun threatening shopkeepers in Victoria Falls, tourists were leaving, hotels were empty.

"Harder times are coming," Matanga said with a world-weary look. Allan rested his head in his hands: "So much damage has been done for no good reason. All I've ever tried to do is bring life back to the land." Then he looked up. "But I guess you shouldn't be here if you can't take a joke."

Late at night the Southern Cross turned slowly on its heel as if to grind the harsh realities of Zimbabwe further into the psyche. Crush the rich, take food from the poor, keep them barefoot, hungry, uninformed, uneducated, and moneyless. That's how a dictator rises to power.

*

At the end of the week Elias Ncube offered to be my guide. "It's well understood here that you can't restore land without addressing social issues, so we've started gender, finance, and health training," Elias told me as we climbed into the Centre's battered old Toyota pickup and headed out. There was a meeting of the village banking women, and we were to join them.

Emelda Ncomo, a bright, single mother of five, was the facilitator for the microfinance project in the villages. Elias had been one of her teachers. Twenty-four banks had been launched. Only women were allowed to get a loan from the fund. The loans were small: one woman borrowed enough for a bag of sugar, which she sold teaspoon by teaspoon. Another bought a chicken and sold the eggs, and another bought a needle and thread.

The meeting was at the "Do-It Yourself Bank." Six women sat on pieces of cardboard in a cement shed. They wore long flowered shirts and bright skirts. A baby was handed around for everyone to admire. Allan had made sure that the headman of each village was on the Centre's board of directors. Chief Shana's wife was one of the bank's members, and she kept a regal pose. Someone read the minutes. They spoke Ndebele, with its clicks and fast rhythms. Accounts were checked and loans were given. "If a woman doesn't pay back her loan on time, she is fined five dollars," Elias whispered. Their briefcases were hand-sewn bags.

One woman complained that the Extension Service came to the villages and told them they shouldn't intercrop trees and flowers with herbs and vegetables, and gave everyone hybrid corn to plant. "All their seeds failed and now people are hungry," she told me.

After, Elias and I were invited to a house for a meal. We sat outside on stools while shallow gourds were filled with a soupy gruel of corn and pumpkin. It didn't feel right to eat the food of hungry people, but Elias insisted that we accept their hospitality. Still, when no one was looking I poured the rest back into the cooking pot so there would be enough for the next meal.

On the way back to Dimbangombe, dust blew. The sun was hot. Five donkey carts flew by, no bridle or reins, just gaunt young men snapping their whips on the animals' meager backs. I asked Elias where they were going. He said, "Nowhere."

*

The next week Allan's thirty-year-old son, Rodger, arrived in camp. He had bought a share of a cattle ranch in Zambia just across the border and invited me to go see it. As soon as I agreed, I received an earful about the problematic "business partner" who had been trying to stage a hostile takeover. Allan looked askance at me: "Be careful," he said.

Rodger and I left Dimbangombe early. Past Victoria Falls,

the dump was smoking. "It's been burning for ten years," Rodger said. Across the border, the town of Livingston was bustling. "All these people are jobless. The only way they can survive is to steal. The whole history of this continent has been twisted by colonial greed over gold and sugar cane. People here were born into a land of chaos, war, and slavery, and yet they are the most peaceful people I know."

In Livingston we stopped at the government offices to pick up Rodger's friends: a policeman and Arthur, one of the Zambian president's aides. They had long wanted to see his ranch. Out in the country the land opened into grassy pastures with wealthy farms on both sides. We passed a large family sitting under a tree. I asked Rodger what they were doing. "Probably nothing. When you have no work and no money, you seek shade."

Rodger explained that his business partner had tried to get him imprisoned. "He and I share assets, but the land is wholly mine. Let's just hope we don't see him there." We turned off onto a well-kept dirt road that led into the ranch. Rodger's mother lived there—Allan's first wife—and I was eager to meet her. Just over a rise, a pickup truck came at us. That's when I saw the barrel of a shotgun peeking out of the passenger-side window, held by a woman. It was Kearny, his partner, and Kearny's daughter.

Rodger slammed on the brakes. "We've been ambushed," he said. Arthur and the policeman jumped out of the truck, then Rodger. He and Kearny stood face-to-face and talked animatedly. Finally, we were left to continue on to the house.

Shirley, Rodger's mother, greeted us. She was red-haired and wizened, her hands shook, and she had the slightly pale look of someone who didn't get out much—an unlikely match for Allan. Rodger put on the teakettle, and just as the water came to a boil, the door burst open: it was Kearny and his daughter, plus four African workers, all pointing shotguns at us.

"I'm making a citizen's arrest," Kearny screamed. He glared at Rodger: "You cunt! Your father Allan fought against everything Africa stands for, and you are just as bad. You're a monkey, a disgusting human being!"

He swiveled around and pointed the gun at me: "You shouldn't hang around such disgusting people. You should be prouder as an American. The Irish built America."

I turned. Rodger was pouring hot water into the cups and keeping his cool, but it was a strategic cool: he was poised to throw boiling water on Kearny. Just then, Arthur stepped forward in front of us to face the intruders: "I'm from the president's office. We don't speak this way. You must be quiet. You mustn't say these things." Kearny's daughter waved her gun at him, then ordered him and the policeman to go out and get into her pickup. Inexplicably, Kearny followed.

As soon as the room emptied out, Rodger whispered, "We have to get out of here!" We ran out to his truck, but the back tires had been slashed, and our food, money, and cameras had been stolen. Rodger had spare tires in the garage, and I helped him change them fast. Then we jumped in and started down the road to the highway. But when I looked behind, there was a large tractor with the four gunmen following us. Rodger turned sharply and we took off over the hayfield, bumping over stubble. "Know how to cut fence?"

"Of course," I said. He told me where to find wire cutters, and as we approached the perimeter fence, I leapt from the moving truck and cut the four strands of barbed wire. Then we bumped up through a ditch onto the highway. When I looked back, the tractor had stopped and was turning around.

Ten miles down the road, Rodger yelled, "Hold on," and we made a quick turn into a farm. "These are friends of mine. We'll hide out here." The entrance was guarded. Dobermans barked and two guards with guns approached, but as soon as they recognized Rodger, we were ushered in. It was a wealthy farmhouse, immense and immaculate. A maid gave us tea and cake behind locked doors, but just before dusk, Rodger decided we should try to make it back across the border to Zimbabwe. "By morning they'll be looking for us and we might not get away this time."

Down the highway we went. Incredibly, the truck broke down three times: twice it was battery cables; then the bumper fell off. I knew quite clearly that this wasn't how I wanted to die. The night was hot and each time we stopped, the people walking along the side of the road stared blankly at us. No one offered to help. There had been talk of thievery and carjackings and I wondered if we would be killed.

Truck fixed, we continued on, but I could see Rodger was thinking about our next move. "I'm afraid they've already bribed everyone at the border to arrest us, so we're going to get some food to give away. In a hungry country, food is as good as money."

We stopped at a curry takeaway place and when we reached the border crossing between Zambia and Zimbabwe, Rodger handed the bags of steaming food to the officers. They looked pleased. He lied to them, speaking Shona: "We've just come from a party and had a lot of food left," Rodger said. "Please enjoy it." They waved us through.

I never thought I'd be so happy returning to Zimbabwe, where life was dangerous enough. Driving at night with armed war vets around wasn't safe. As we gained altitude, the night air felt cool and I sat back, trying to relax, trying to come to grips with the irony of having almost been killed by a white racist in Africa whose hatred was for a white man who was a war hero of the black Africans.

A few hours later we pulled into camp. It was two in the morning, but Allan was still up. He tried to appear unconcerned.

"You must have had a hell of a party over there," he began. We told him what had happened, about our escape. I commended Rodger on his cool behavior. Allan teased me gently: "Weren't you having enough fun here?" he asked. I felt euphoric and exhausted. Allan gave me a hug. The gun cabinet by his bedroll was open. I went to my tent and fell asleep with the sound of baboons rustling in their treetop nests around me.

In the morning we took it slow. It was Sunday and Allan debriefed his son. A car pulled up. It was Arthur, the president's assistant. In a shaky voice he told us what had happened: "Those four men with the shotguns—I told them who I was and gave them some money and they left. The daughter took us to a police station, where the officers had been bribed to arrest Rodger. They kept us there all night. But when we gave them more money than Kearny had given them, they backed off."

Allan poked at the campfire and we brought him a plate of fresh fruit. He said, "I've never heard people talk like that man did. It is shameful. It is still hurting inside my ears."

It was clear that Arthur had saved us. We would have been easy to find, bumping along the one highway. We offered more fruit, then rusks, coffee, and chocolate. He looked around at our camp admiringly. "I'd like to live this way," he said.

Allan suggested to Arthur that he hunt an impala and take all the meat home. That's how you say thanks in the bush in Africa, and we watched as Arthur crossed the river and went off through the thorn trees to look for game with a borrowed gun.

12.

In his journal, Emerson wrote: "Everything teaches transition, transference, metamorphosis. . . . We dive & reappear in new places."

At latitude 81° north, longitude 30° west, it was just past ten in the morning on the first day of July 2001 when I hoisted a sixty-pound pack on my back, more than half my weight. In the company of six other seasoned northern travelers and the photographer David McLain, I followed Dennis Schmitt across Warming Land, a triangular peninsula on the northwest tip of Greenland, five hundred miles north of the northernmost inhabited village in the world, and five hundred miles south of the North Pole.

Warming Land was named after a Danish botanist, though at that latitude, hardly anything grows. The trip would last three to four weeks—the entirety of what passes for summer.

"Only glaciers have walked here," Dennis said. His hiking boots were old and only half-laced, and a pair of ski goggles hung on stretched elastic around his neck. "We are going to a land where, as far as I know, no human has ever set foot. You will base your well-being on each other, and we will travel as a miniature society for three or four weeks in what will seem like a small lifetime."

Sandy-haired, paunchy, disheveled, Dennis had a genius IQ and an irreverent restlessness. He'd divided his life between Arctic exploration and music composition. He had left home at seventeen and lived with a widow at Anaktuvuk Pass in Alaska, learned Inuktitut, and began making first ascents of Arctic mountains, including the 13,176 foot-high Mount Marcus Baker in the Chugach Mountains plus others in the Brooks Range. He returned home to get his degree in anthropology and linguistics at Cal Berkeley, and on the side, he composed classical music.

The Arctic kept calling—a kind of polar madness propelled him. "My focus had been Arctic Canada. I climbed on Baffin Island and northern Ellesmere Island, and as soon as northern Greenland was opened to outsiders in 1995, I immediately explored Wulff Land, Oodaaq Island, Kaffeklubben Island, and Cape Morris Jesup, where I climbed the most northerly mountain in the world." From these weather-layered ascents, Dennis wrote contemporary symphonies whose movements replicated the movements of his feet up and down remote unnamed mountains.

On the flight north from Resolute, Nunavut, in a fat-tired Twin Otter, the two pilots clipped an aeronautical map between them that Dennis had already calculated was off by eleven miles. Northern Greenland had not yet been mapped. "We navigate the old-fashioned way," one of the pilots said, smiling and pointing to the sextant bolted onto the top of the instrument panel.

The Greenland-born explorer and ethnographer Knud Rasmussen wrote in his notes during his near-fatal 1917 expedition to the top of Greenland: "Quite near us we saw St. George's Fjord, narrow as a river of ice cutting into the land, encircled by high mountains. . . . There is a breadth here and a depth, a wild mountain grandeur. "

Eighty-four Julys later, our group began a journey across what Dennis described as "an Arctic Yosemite," with dihedral buttresses, polar glaciers, and fast-moving rivers. Hunger was the greatest challenge for Rasmussen; I wondered if it would be the same for us. Before we left Resolute, Dennis suggested we each eat a whole rasher of bacon. "You'll need the fat," he said. And we did.

On our six-hour flight northeast, Dennis lurched from side to side, pointing out places where he had wandered. On the left side of the plane, Ellesmere Island was scalloped mountains, giant snowfields, blue-blasted crevasses, and fluted channels of ice. On the right, Lake Hazen sparkled. Under us, the Kennedy Channel was choked with upended slabs of pressure ice and pocked with decapitated icebergs and scuds of drifting snow. Hall Basin's L-shaped beach was crowded with pancake ice. The cracks between platelets had already begun freezing, reminding us that "summer weather" at this latitude was brief—perhaps no more than twenty days.

After a long, airborne reconnoiter through narrow canyons and over glaciers, we landed on a wide plain near Saint George Fjord. I stood in place and did a three-sixty: down-sweeping tawny cliffs, bright ice tongues between what looked like brown buttocks, tight-joined limestone underfoot, calved ice drifting toward the channel. An all-encompassing silence descended. "This is Warming Land," Dennis whispered in my ear. We pitched tents, boiled water, drank soup and tea. All night the fjord held sun like a silver knife; the top of Hendrik Island was a gold hat. It was there that the Inuit hunter Hendrik Olsen had disappeared. He was Knud Rasmussen's favorite team member, the best hunter of the group. He had gone to the island to get a seal and was never seen again.

In the morning I heard Dennis's sweet voice: "It's time. Don't worry, we're just going for a little walk." We hoisted our packs. Underfoot, calcified limestone made a firm floor. Farther on, frost-heaved rock erupted into patterns: sorted polygons and circles of "fines," smaller rocks surrounded by larger cobblestones as if a mosaic-maker had been through. Instead of traveling the sea ice by dogsled, we made our way on foot over felsenmeer, "sea of rocks." For navigation we had only a satellite photograph taken from twenty-eight thousand feet. There were frigid rivers to cross and glaciers. We practiced self-arrests with ice axes; our crampons were tied onto the tops of our packs.

The first day we leapt across a five-foot-wide gorge of roiling water and trudged over a mountain topped with "quickmud," sinking to our knees in thawing lobes of earth. Where mud mixed with ice, we glissaded. Life at this latitude was almost nonexistent, except for the redpoll we saw atop a boulder. Brown with a red cap and a pink breast, this bird survives the Arctic winters by storing seed in its esophageal pocket and taking shelter in lemming burrows.

To step on land that perhaps had never felt a human foot—I had thought that was a privilege for those going to the moon. From the notes Rasmussen made during his Second Thule Expedition, I couldn't be sure that he hadn't traversed Warming Land on his way home from his disastrous, hunger-filled journey. That made this arduous trip even better for me: to think that the man from whom I'd learned the most about Arctic culture had been there too and that, perhaps, our paths were overlapping.

While taking a drink of water from a roaring glacier-fed river, I fingered a piece of fossilized coral. This had once been a shallow sea. Now it was a polar desert. Arctic willow was splayed out flat on the ground as if trellised there. Nothing grew more than two inches off the ground.

I spied a woolly bear caterpillar basking in the sun. This type of caterpillar grows so slowly it can take fourteen years to develop. The woolly bear caterpillar freezes solid in the winter, finally pupating and emerging as a butterfly, then lives so hurriedly, it doesn't even bother to eat; it simply has sex, and as soon as there's a hard frost, it dies.

We came into a long U-shaped valley. Cliffs swept down, tawny and red, and a fast-flowing river pooled on marble slabs, spun, and fell. Everywhere we could see how ice had made this place. Glacier ice had scratched, scoured, abraded, and polished every surface and hollowed out whole valleys, cut deep lakes. The canyon deepened as we walked. On one side a glacier's crumbling edge broke into turquoise spears.

I stopped once to rest and looked up: above was a hanging valley with its own tiny ice cap melting into a wispy waterfall. Ahead was a half dome—the counterpart of El Capitan but made of limestone, not granite. "This is what Yosemite must have looked like during the ice age," Dennis said admiringly. He slumped down against a rock, furiously scribbling a few bars of music, then said: "This is the place I feel most at home."

Every day was a long hike up and over mountains and ice. When we crossed wide frigid rivers, we took off our boots and pulled on neoprene "socks" worn with Teva sandals. Before stepping into the water, we unstrapped our backpacks in case we fell so the weight of them wouldn't cause us to drown. But on the other hand, if we lost our gear and food, we'd die anyway, so we didn't let it happen. We linked arms or used our trekking poles for stability, and slowly made our way across with our pants pulled up over the knee—anything to keep our clothes dry. Sun out. No wind. "It's the kind of silence in which you can hear the rocks getting old," Chuck, one of our group, said. We followed the edge of a lake. During a rest I took off my boots and wiggled my raw toes, then doctored the abrasions and blisters. We heard water falling stereophonically: the remnants of the last ice age melting. A hundred tiny streams wormed their way through gravel; where water leaked down the two-thousand-foot-high scarp, the rock face was black and gold. "Here we are at the end of time," I said to no one in particular.

Four days in and I was already feeling hunger. I'd come from a high-protein meat diet in Africa to one that was only candy bars and dried soup. I had packets of crackers and cookies and one sausage that I rationed. One slice a day—that's all I permitted myself. As we walked the deep glacial trench, we saw no scat, no spoor, no animals. We were the only megafauna around.

Thick mats of saxifrage covered the rocks. *Saxifrage* is derived from the Latin for "rock-breaking herb"; the plant has an uncanny power to burst through the calcium-rich limestone latticework. Purple saxifrage evolved in the high Arctic. Its buds overwinter, so as soon as the snow melts in June, flowering can occur. Three or four weeks later, the ripened seeds cross-fertilize if there is a bumblebee or a butterfly around. If not, the male anthers grow long, bend inward, and as soon as a gust of wind pushes them toward the female stigma, pollination occurs. When purple saxifrage is in bloom, elsewhere, on the southern part of Ellesmere Island, caribou are calving. The entire season of one plant is thirty days.

Thinking of those botanical adaptations, I regretted how ill prepared we humans are. Survival was on our minds—not that I was frightened of our deep isolation. I was simply vigilant. But I rationed my food too earnestly and discovered that the salami I'd brought had gone bad.

Everything solid that I'd been able to carry—crackers, cookies, candy bars—was gone by the end of the second week. Hunger woke me: I stood outside my tent and licked the pockets of my parka for crumbs. Nights were below freezing. Days were barely above. It was always light.

Hunger became an ally. My metabolism changed and my understanding of this land changed with it. On the night the wind howled, our tents rattled like bones. We were camped by a string lake. Pans of ice made of bunched crystals floated by. Pale green on top, the clear sides looked like see-through rows of teeth. When the sun came, the bunched stalks disintegrated: deconstructed chandeliers. I heard music—not Dennis's but candle-ice tinkling. The whole lake chimed.

Lying on top of my sleeping bag by the water, I lost track of my body. I wasn't floating—there was nothing mysterious going on—but something had let go inside me. The weight of my boots, my abraded heels, ankles, and toes ceased to hurt and no longer impeded my journey. I had entered a trance state. The equation was this: hunger + beauty = movement. I wanted only to keep going.

Some days we crossed musk ox tracks, but they were old. There were no seals in the fjords, no fish in the lakes or tarns. It was easy to see why Rasmussen's crew starved. Boundaries gone, an "anything's possible" mindset took over. References to what was or what could be seemed confining. I had come unfastened.

Every day I felt stronger. Every day I lost one pound. Dennis showed me the rumpled pages of his new symphony's first movement and hummed it directly into my ear as we walked. The music was a wild continuum that matched this place: icy and percussive with bold, strident progressions followed by melting harmonics. At the same time, I began hallucinating. Sprays of light shot out from the ends of my fingers, and each time my feet pressed down, a chord sounded. The earth was an instrument, and my walking on it made music that only I could hear.

Exhausted and cold, my eyes were drenched; I felt ecstatic. Along the way, Dennis and I stopped to stare at a revetment.

High up on a wall was a huge cave. "Big enough to live in," Dennis said. "Big enough for a Steinway?" Chuck asked, teasing, but Dennis didn't hear. He was already shuffling through and marking his battered sheaves of music.

A waterfall pulsed off a wide lip of rock rounded by thousands of years of tumbling. For days we'd been walking uphill. Now we were on the other side of a mountain and the vegetation became even more sparse. Rock prevailed. "I've been surprised by its uniformity," Dennis said. "It's a monoculture of limestone, an old seabed through which a glacier recently passed."

Without knowing it, we had entered the protected midsection of a canyon, a platform teetering on the divide. The canyon deepened as we walked alongside a river running by one wall. It was leading us, but I couldn't have imagined where. Then we entered a great "hall." An enormous flat spread out before us, a plain of rock. A single glacial erratic big as a house lay in the middle, frost covered and cracked jaggedly in half. Meltwater streams poured from the toes of the glacier on the far side and threaded through the gravelly till, ending in tiny pools that shimmered.

The glacier itself was a rough wall of collapsing pillars chiseled by sun. A piece fell, echoed, went silent. David walked up beside me, his cameras finally at rest. "This is the courtyard of the gods," he whispered. We were silent for a long time. If ice has a memory, would it ever forget this place?

"Up here you are looking at a geological landscape, the same rocks that were here millions of years ago, but rearranged by glaciers," Dennis said. "Where we are right now has nothing to do with human time. The word *now* is meaningless. What we call a year is a tiny framework in a huge sea of time. We are engulfed."

I lifted my arms. No words came, only images of the Japanese gardens I had once visited: Saihō-ji, Kinkaku-ji, Ryōan-ji. But this place was the font. The whole world was this, embedded in this, had issued from this. It was a place where, as Dōgen said, being and nonbeing are rolled together.

Fresh, original, unvarnished, spontaneous, and old. No hinge that held or released. Some places remind you of nothing and everything at once. That's what this basin did. We were suspended somewhere between Nyboe Land and the Sherard Osborn Fjord as if suspended by one of Dennis's musical interludes.

I stepped forward. Would I fall? Water flowed around the toe of my boot, as with the glacier that leaked meltwater from its foot. The basin had a Japanese severity. It was distinct, direct, precise. Yet everything seemed to overlap everything else: crumbling glacier ice, meltwater splashing against pincushions of moss, moss edged by shoaling gravels.

Time to leave. It wasn't silence but absence descending as I stepped back. No grief, nothing tender, only this hard hall of light and rock. I was tired. We had walked almost the entire unmapped peninsula. At the last moment I turned to bow.

The width of Warming Land is just under forty miles, but the topography is complex. Our only understanding of it came from Dennis's satellite map. Our feet translated the intellectual "above view" into a blistered, footsore reality. We were often "rimrocked"—unable to get through a mountain pass—and had to turn around and try another way.

Perhaps Knud Rasmussen had seen this place—I hoped he had. But no other human had ever been here, had breathed its crisp air. One very high-altitude airplane—perhaps a spy plane—flew over each night. That's all. We walked in light, slept in light, walked in light. Each step took us north toward the next fjord.

"Because we go, beauty stays," the poet Joseph Brodsky wrote. It was getting cold and we had to make camp. After an hour's walk up-valley, we set our tents by a narrow lake, where we stood for a long time bundled in parkas watching lake ice move. I slept with the tent flap opened and my head pushed outside. A lone glaucous gull, one that we had seen briefly the day before, had followed us there, hoping for food. I had none to give.

I longed for a day of rest, but "mad" Dennis had other ideas. More jittery, obsessed, and withdrawn than ever, he announced that we would be going on a twenty-four-hour-long hike to Sherard Osborn Fjord. We would begin at midnight. "So get some sleep now," he said cheerfully. When he saw our looks of disbelief, he cajoled us: "It's only a walk, just a walk. . . ."

A roar woke me. "Avalanche!" David yelled. I dragged myself out of the tent: a piece of the ice cap had fallen, burst midair, and tumbled down but stopped well before it reached us. Later more curds of ice crashed down the rock face. At seven p.m. it rained.

We slept until ten in the evening. Dennis woke us to get ready to go. We ate a snack, then marched off with only our daypacks. Although relatively unencumbered, we were so tired and our feet and ankles were so raw, the feeling of liberation was hardly noticeable. When the temperature dropped suddenly, snow squalls roared over our heads.

Up and over the mountains we went, past entrances to side canyons. There were five river crossings, slow traverses through quickmud and over scree slopes. I no longer tried to avoid pain but slammed my feet down hard, using my spine as a shock absorber. Someone found an Arctic hare's foot on the ground.

"Should we bring it for good luck?" Chuck asked. "We may need it."

Mist wound around a spire of igneous rock. With an eye out for geological changes, Dennis quipped, "It's an anomaly, like me." Another whole canyon system appeared out of the clouds, and its rushing river merged with the one we were following. Curtains of shadow moved across canyon walls as if pushed by an inextinguishable sun.

Along the way David confided that he wasn't feeling well. "Every time I turn my head, the world spins," he said. Dennis was so far ahead we couldn't catch up to tell him we had a problem. His mountain-climbing passion was opulent and out of control.

Helping David, we climbed down the side of a waterfall and up an almost vertical slope. Buttresses flew by. Thousands of feet below, a deep gorge was filled with splashing water and boulders. It was six in the morning when something sticking out of the ground stopped us. A bone, possibly a human femur, surrounded by three flat rocks. Was this a cemetery? I looked for skulls, more skeletal remains, but found nothing. The bone was old—how old? "It may not be human," Miki said. We listened to her because she was a microbiologist. "I don't want to disturb it," she said, and replaced the bone on the ground.

The slope opened up and we entered what looked like a Roman amphitheater: before us were massive blocks of terraced rock in long rows like theater seats, all facing the engorged river below. "Those Romans really got around, didn't they?" Frank, Miki's husband said, and yelled out a few lines in Latin. The words echoed. "Yes, they had to be here!" he said, laughing.

David's dizziness worsened and he sat on the scree. Below was a vertiginous drop to the river. One bad step and you could plunge. We could see where Sherard Osborn Fjord began, and the monolith Dennis wanted to climb—it was just around the bend—but Miki took refuge with David and announced that she couldn't go any farther. She had started out as the strongest; now she said she was frightened. David's head was hot. Dennis and Bill, a journalist based in Moscow, had gone on. After a long discussion among Chuck, Frank, Miki, and me, we turned back, one of us on each side of David until, closer to camp, the way flattened out and we could see our tents.

We had been walking for fourteen hours. The pace had slowed to a shuffle. When we finally arrived at camp, we ate a light meal of soup, raisins, and tea. Snow came, and soon our whole world was white—the lion-colored cliffs and the long lake. My yellow three-season tent was the only sun. Dennis had said to wait twenty-four hours before worrying about him, but that was before the storm began.

At some point I crawled out, too oblivious to realize I wasn't wearing any waterproof gear, and I wandered. Up on a gravel ridge I looked in the direction of the fjord. Would Dennis and Bill find their way back? Would David get better?

The snow was wet and slapped my face. After years of travel out on the ice with Greenland subsistence hunters, I knew how quickly things could turn bad in the Arctic, and a feeling of dread filled me. The stupid, joyful drudgery of the trip had become a daily ritual—walking across the top of the earth on glacial rubble—but how pointless it suddenly seemed.

Rasmussen had gone hungry when he passed through Warming Land. He had to eat most of his dogs, and his botanist, Thorild Wulff, starved to death and was left behind. Was that whose bones we had discovered? Chuck, Frank, Miki, and I had a meeting about what we would do if Dennis and Bill failed to reappear and if David grew sicker. Could we find our way back to where the plane had left us off? Could we carry David?

Voices woke me. It was Dennis and Bill. Thirty-six hours had passed. "It was difficult and thankless," Dennis said, laughing. "The snowstorm hit when I was on the summit. The front of the monolith was like a diving board—three thousand feet straight down. In the whiteout, I made sure I turned the right way before starting out."

Bill said, "We were pretty tired. At one point near the bottom, Dennis lay down to drink from the river and dozed off with his head in the water!" To which Dennis replied, "I can sleep anywhere!" Then he dived into his tent to sleep, goggles still on. After a long rest and something to eat, we all started for "home."

Backpacks lightened, feet bandaged, and inured to pain, we were like horses trotting back to the barn. A dusting of new snow lay on brown palisades. Across a wide valley, alluvial fans aproned out into meltwater deltas chinked with mud the color of deerskin hides. At midnight all but one cloud vanished. The air felt warm and the lake where we camped turned from cerulean to celadon—fine porcelain you could see through as if into the true nature of things.

We slept by a stream. I didn't bother with the tent. Rock was my bed; walls were made of ice; the roof was changing weather. The snow changed to bright sun and back again. David's whirling world had slowed. The pace was leisurely, and between river crossings we took wolf naps. The crane fly that had landed on my notebook the week before and walked between words returned. The glaucous gull didn't. Only one of its feathers remained, stuck between two rocks.

From the river we climbed two thousand feet straight up, only to find that there wasn't a way to get to the next ridge, so we climbed back down and went around. Before us appeared a steep-sided, squared-off lake walled by an icy terminus that was calving green leaves of ice. We teetered around the edge, then scrambled up a rock face before strapping on crampons and traversing a badly deformed glacier where a fringe of bent icicles hung from its side—proof that the glacier was moving.

At the snout I leaned in to listen. "Can you hear it breathe?" Dennis asked. I nodded: "This is our only keeper of time." We put our ears to the meltwater ticking. It was summer's fast clock in a slow geological world.

Beyond the lake we entered a narrow pass of linked tarns. Alpine poppies flapped in the breeze. We stopped to rest in what seemed more like Switzerland than Greenland. The plants were alpine. Chuck, who had carried extra weight for me the entire trip, curled up and napped on a bed of shale—ancient layers of mud that felt much softer than the previous seventy miles of sharp limestone.

A single bumblebee flitted by. The *Bombus polaris* generates heat by shivering. Only the queen survives the winter. In the spring she hatches out a new colony of workers and starts over again.

The last day of walking was long. For every downhill, there were ten uphills. "Is that mathematically possible?" Miki asked, groaning. She was limping now from a sore Achilles tendon. I thought of the great warrior, Achilles. Miki said she must have been wounded in the heel by Paris's arrow. We were limping and laughing until finally we walked onto the much-anticipated "runway" where we hoped the Twin Otter would land.

The temperature dropped and we ran up and down the tracks the plane had left weeks before to warm ourselves. Dennis pulled the satellite phone from his pack and called Bradley Air in Resolute. He gave the pilots a concise weather report: "Two-thousand-foot ceiling. Precip starting. Wind from the west at five knots." The pilots said, "Can't come. Call tomorrow."

Dennis paced back and forth humming newly composed bits of his symphony's first movement, then gave an impromptu disquisition on modern music. "I've been using the fugue but taking it further," he explained. "The emotional and imaginative effect is more important than sticking to a rule. But when it threatens to become noise, I eliminate it. Too much twentieth-century music is based on a mathematical idea that turned out to be unlistenable. Too much density and the human ear can't stand it. I used the twelve-tone system to create fragmentary effects, like adding a spice—a context for a return to tonality."

That night we boiled water as usual to make our "add-water" suppers. I was hungry but found I couldn't eat anything solid—so I gave the few dehydrated meals I'd brought to David. Instead I drank broth and tea and tried to imagine the agony of those who had actually starved up here.

Bad weather closed in and the ceiling dropped from two thousand to five hundred feet. Dennis kept his calls to Resolute short. He had to conserve the batteries. When he called at eight in the morning, the pilots said, "Can't make it. Call tomorrow," and we began to wonder whether they would be able to come for us at all.

Day three. The pilot said: "Ceiling too low. Don't know when or if we'll be able to get you." After, Dennis stood before us, ever more disheveled but intensely alert. He ordered everyone to bring out all their food and spread it on the ground so we could calculate how many days we would be able to survive. He looked at the meager array and shook his head. "There must be more," he said, and burst into the tent of the "quiet couple," the ones who stayed aloof from the rest of us. He came out holding two stuff sacks full of food: a wheel of cheese, sausages, candy, dehydrated meals. He looked at the couple, furious. "This belongs to all of us!" he said. "And when we get truly short of food, you will not be given any."

No one spoke. We boiled water and stirred our meager soups. Hard to sleep that night. The sun shone on our hunger. On the fourth day, when Dennis called Resolute, a woman answered the phone. "They left at six this morning. Not sure if they can make it. Can't refuel at Alert, so bringing fuel barrels." Click.

We held vigil by the side of the runway all day and into the night as heavy clouds—the ceiling—lowered to the ground. It was cold. With no sun, we were losing hope. I remembered how the two pilots grinned and shook their heads when they left us at the beginning of July. Now my clothes hung on me, and with no hairbrush, my long hair was tangled beyond repair.

We napped in our sleeping bags enshrouded by a cloud. Then we heard a distant noise. It was difficult to know what it was or where it was coming from. We stood silently, backs straight as if at attention. Dennis held his hands behind his ears, then smiled. "It's them."

The Twin Otter lifted straight up out of the mist that hung in Saint George Fjord like some monster, made a quarter turn, came straight for us, and landed short on its fat tires. The pilots stepped out. "We did a scud run all the way from Resolute. Couldn't stop at Alert. Weather bad there. Have to refuel here. We'll hand down the barrels."

Ten heavy barrels came down. Hose attached, we took turns hand-pumping fuel into the plane's tanks over the next few bone-chilling hours, then reloaded the empty barrels back onto the plane. Strapped in and shivering, the pilot turned and said, "Sorry, can't turn the heat on. Too many fumes." We wrapped ourselves in sleeping bags and parkas as the plane took off over the frozen channel below.

I was shaking with cold when Dennis wrapped an extra parka around me and hummed the music he had been writing into my ear. We flew over Petermann Glacier, which would soon calve off a block of ice bigger than Manhattan; followed the uninhabited coast of Ellesmere Island; and crossed Nunavut's eastern archipelago, including Beechey Island, where the remains of Franklin's starving crew had been found. Dennis had asked the pilots if they would call ahead to Resolute and ask someone to open the kitchen and cook something for us. Anything would do.

All the way to Resolute, Dennis looked back at the country behind us, at the ice-capped triangular lobe that was Warming Land, and Hendrik Island. It was one of the many fingers at the top of Greenland that grasped the last of the earth's ice.

"It's not easy to leave," Dennis said to me. "You could say that polar desolation is my joy. It's where my music comes from." Then he leaned back and closed his eyes. "Ice is the god of this place. It's here that we refresh our souls."

13.

Perpetual motion and the banality of it. Or the thrill of staying in one place. I had been in my new cabin in Cora for a month. A moose and her calf ran by and I trotted behind them, then crawled on hands and knees to peer down at the pond where they stood knee-deep eating wet grass. When the flies descended, the moose kneeled down until water covered her back. All wet, she stood and shook, called her calf, and made her way to the next kettle pond not far away.

I spent the rest of the fall alone. Just me and my kelpie, Gabby. My favorite dog, Sammy, died when he was seventeen years old. That last week of his life, his hearing came back and his vitality. He stood up in his cage at the vet clinic where he lay dying and barked. The bark said, "I want to get out of here." He seemed suddenly well, and I took him home. A week later, he was dead.

I would always regret the time spent away from him and longed for the deep, quiet pleasure of his presence. Who would teach me about unconditional living? Who would tell me that staying home was the same as leaving? After his death, all the red and orange aspen leaves fell in one night, and the myth of freedom was like a million gnats nibbling my ears.

When my new cabin was finished, Gabby and I moved in.

The next day I left her there when I had to go to town, and she escaped: I spent two days looking for her. Finally, the housekeeper at the Kendalls' ranch, where I had been staying, called and said, "Your dog is here in the cabin. She's waiting for you." Gabby had walked the seven miles back to the ranch she'd thought was hers.

I needed a well and the drillers finally found time to come. The night before I'd cut two lengths of copper welding rod, bent handles at the end, and walked out into the pasture, arms straight out, hunting for "the pluck of water," as the poet Seamus Heaney called it.

Water is secret and sacred. We pay little attention to its underground sinuosity, its spills and ruptures, its entanglement with geology. Here, it was granite and the well men said they would have to bring a diamond drill.

In places the rods shuddered slightly; then they suddenly crossed against my chest. I hadn't moved my arms—the copper rods forced them to make an X. "The pluck came sharp as a sting," Heaney wrote in "The Diviner." I walked forward, then back, and the same closure happened. I marked the spot with a post. "Drill here," I told the men the next day, and they did as Gabby and I watched. Mud and sand came up, some coarse, some smooth. When they added two more lengths of pipe, water came: fifteen gallons a minute.

After, I hung the rods up in a safe place and thanked them. Water is how time plays its notes, announcing mortal wounds, unvoiced regrets, unexpected spurts of joy. It is time's unbroken medium; it carries us. Water is alive; it remembers how it has been treated.

Winter arrived, and when the snow neared two feet on the level, I packed the pickup, loaded Gabby into it, and drove to California, where I had a house on the central coast. I stopped one night south of Ely, Nevada, at a campground with a horse corral and a tiny creek winding by. My cowboy bedroll was in the back of the truck and Gabby lay close against me. It wasn't the same without Sam. We three had traveled many miles, back and forth, cross-country, south to New Mexico, north to Montana, on horseback, on foot, in the truck.

By the next evening at dusk, Gabby and I had reached Gaviota. On the road to my house I said the word "beach" and Gabby tilted her head, then exploded in happiness. She and I ran for miles across shining aprons of wet sand.

At Christmas, terrorist threats lay under every tree, and the holidays rolled downhill in muddy ruts. The new year opened mournfully. Fog elbowed in; then rain fell, washing away what had not yet begun. When the rains stopped, drought took a firm hold and I saw that the moving marble of sky was only a confection.

Gabby and I walked the hills, past eroded sandstone outcrops and along the ridge where manzanitas' smooth arms were entwined. She stayed close. Any strange sound kicked my lightning-induced hypervigilance into gear, and the memory of the Twin Towers attacks kept it kindled. Terror had entered our lives and was there to stay. I dreamed that rows and rows of planes, four abreast on approach, were shot down. I was captured and led to a small room where there was a rough desk and a small light with a tin shade and told, "This is where you must write from now on."

*

Very little rain fell in the rainy season, and soon the chaparral—white sage, black sage, and scrub oak—went brown and gray. Live oaks tipped over dead; acorns were empty shells. There were no mats of spring wildflowers, and the grasses were separated by bare ground. Cattle were sold, and because the foreman refused to implement a strict program of planned grazing, the soil surface of the ranch's fourteen thousand acres was hard, making the little rain that came more likely to run off than sink in.

The ice age that should have been arriving, according to the Milankovitch cycles, faded under the uphill signal of global heat. Gabby waited for me to utter the word "beach" so we could go to the ocean and cool down. When Sam was alive, mornings were calmer: he'd bury his dog biscuit and leave it for Gabby. She'd wait, haughty and entitled, watch where he nosed it under the dirt, then, after a few theatrical beats, dig it up and eat it while he watched.

Now, in Sam's absence, she took the biscuit, buried it herself, and never returned for it later. "Solace doesn't arrive on a silver platter," she seemed to be saying. Sometimes it doesn't arrive at all. It can hide under dirt or wet sand, though sometimes a bright edge will show and stir our private and public agonies momentarily out of sight.

14.

The moon shone on stacked thatching grass and a hyena cried. It was May 2002 and I'd returned to Africa. Allan greeted me with his usual laughter and sardonic quips, and we drank tea around the campfire. Out of the blue he recounted the night long ago with a friend around just such a fire when a lion jumped in, grabbed his friend by the neck, and dragged him away.

Allan used the story to inform me that things in Zimbabwe were much worse since the last time I was there. A year had elapsed. Now people were truly hungry, and half the staff at the Africa Centre had died of AIDS. Many of the trees that lined the road to Dimbangombe had been cut down for firewood by outsiders. Across the highway, the villagers' livestock was still allowed to run loose, elephants marauded through gardens, and the land remained predictably bare. By comparison, planned grazing at Dimbangombe had produced thick grass.

Mugabe's land resettlement program had become dangerous and chaotic: his cronies were being given land as rewards for loyalty, and they were finding life in the bush harder than war. But Allan saw it as an opportunity to stop poaching. "In the Mau Mau wars we learned to turn our enemies into allies," he explained. Following suit, he hired a well-known poacher to become an anti-poacher and keep an eye out for others killing animals illegally.

I kept returning to Africa because Allan was part of the threadwork that taught and held me to the life of the open range, the operations of photosynthesis, and the possibility of reversing climate change by putting airborne carbon into the ground. I loved being with him there, and what better place to see firsthand how degraded and desertified land could be restored.

A piebald kingfisher flew up and down the stream by our tents looking for fish. "Everyone is hungry," Allan said, watching the bird, "except here, where we grow our own food." Zimbabwe was experiencing a terrible dry spell, but everyone at the Centre had a vegetable garden, chickens, goats, and part ownership of cattle. "There's no such thing as drought!" he exclaimed, laughing. "Bare ground causes drought, then drought causes more bare ground. But if you let grasses loosen the soil to let moisture in, things will grow."

After breakfast the first morning, I went with Allan and Matanga to see a new communal garden. "Five of the villages are now competing to see who can grow the most food," Matanga said. "And that's good. They are learning about harvesting water, making compost, and saving seeds. They are coming to see that nature's intelligence is much vaster, more diverse, and complex than ours."

In other villages things were bad. One woman said, "This has been a hard season. No water to make a garden. No garden at all. And we have only one borehole for the village. We can't buy any mealie meal [corn], so we eat millet, but the birds ate most of it. Some got UN food. We didn't because they know we oppose Mugabe. I make crochet hats to sell while we wait for the rainy season to come. My family is all right, but AIDS is killing many."

At dinner I asked Elias if he would again be my translator and guide. I wanted to see more of the villages. How would they deal with the rainless months ahead? As we bumped along the dirt road, Elias said his father "read the weather" and could treat snakebites and other ailments. "He pounded the brown seedpod of the umhlanziso tree into a powder, and it was given to prevent malaria. But there is no medicine for AIDS."

Clouds gathered, and the mentally ill woman who had been living by the side of the highway had hung all her clothes on a tree and lay on her back in the bush, half-naked. A donkey cart galloped by, the animals beaten mercilessly by whip-yielding boys. We stopped where there was a funeral going on. A tall, thin Catholic priest waved the censer over a coffin. "In the old days, they'd put a knife in the dead man's hand," Elias whispered. "He's called a *suntwe*—a hyena. That animal is associated with death. It's also a witch's animal. You are to make a sound like a hyena during the ceremony, and the chief gravedigger makes hyena sounds to frighten the spirits away. But as the elders die, the old ways are disappearing. Now they do a Christian burial in the morning and the traditional one at night."

A donkey cart passed us full of children singing. We stopped to give a woman a ride. She said, "There is actual hunger here now. Some days we eat only watermelon. Our corn died." An oxcart filled with furniture went the other way. Elias looked: "Someone is moving. That means someone has died. When parents die of AIDS, the children are taken to relatives. Some have many children and there's not enough money for school fees or food. But some children are left with no extended family at all."

The land was puce and ocher and covered with rocks. At a house we visited, someone was milling corn with a hand grinder. We were offered soup made of corn and pumpkin but it was more watery than the year before. A young girl ironed the skirt Mrs. Ncube would wear to town for a meeting using a nineteenth-century iron heated with bits of coal.

It was hot. Goats and cattle wandered loose. We drove back to Dimbangombe, where a gathering of Allan's students were talking about drought. "Down in the low velt, dryness can be worse than here," Chris said. He was one of the educators being trained by Allan. "In the 1960s the warthogs stayed alive by eating the bones of dead animals. I had two horses and could barely keep them alive. At the end of each day, I'd collect buckets and pick green grass by the edge of the highway for them. When the dry lasted for two seasons, we lost sixty thousand impala. The Sahelians say there is no such thing as a one-year drought. Animals gain weight in the wet season and lose it when it gets dry. But the second year, they lose weight all year round, then die."

Allan listened, puffing on his pipe: "You should plan every year as if it's a drought and always keep a drought reserve. That way, if there are good years, you have a bonus!" Before dusk we walked to the edge of the vlei. It was alive with guinea fowl, impala kicking their heels, pairs of baboons grooming each other, and shy kudu hiding in the bush. "When I see this, I'm happy," Allan said.

That night the new moon stood straight up like a spear with planets top and bottom. Clouds gathered, but a strong wind carried them away. The next morning, we walked with the herders and eight hundred head of cattle through acres and acres of waist-high grass. Allan said, "I wish that people would come and see that even when there's no rain, we still have healthy cattle. There would be no drought if the soil was taken care of, if rain was allowed to seep in."

The next afternoon I accompanied Allan to a meeting with the ZANU-PF soldiers who had been put on land at the edge of the Africa Centre. When we drove up, everything grew tense: the men grabbed their guns and walked toward us. Barefoot and unarmed, Allan stepped out of the jeep to greet them. I stayed behind.

They sat in the shade of a big tree and listened as Allan explained his offer of jobs as anti-poachers. They would get a uniform, a gun, shoes, meat, and a salary. One or two agreed. The others sat stone-faced holding their rifles. When we left, I wondered if we would be shot in the back, but we arrived at camp safely.

The following day Allan had to leave for a meeting in Bulawayo and left a loaded shotgun on my cot with a note: "Use only on humans, not animals. Back tomorrow." Lions roared all night, but no angry soldiers came down the road. In the morning when Allan returned, we watched a kudu bull drink from the stream. "Animals know more than we do," he said. "A male giraffe that had been pushed out of the herd went to live many miles away. But when the leader of that harem died, the giraffe somehow knew there was an opening for him, and he traveled a hundred miles to take over the herd."

Later I helped feed Allan's adopted baby elephant, whose mother had been poached and killed. The baby wrapped her trunk around my arm and looked into my eyes as I gave her a bottle. "Elephants are very sensitive, very nurturing," Allan said. "They bash through the bush, pushing down trees and eating all the fruit, but they care for each other beautifully. When they get nervous, they hold one another's tail."

*

Hot, rainless days continued, and the following weekend, Elias and I drove back to the communal lands. I'd heard about a "rainmaker," and I wanted to meet her. The talk about drought had made me wonder what the traditional people thought. I wanted to learn about what they called "the spirit side of weather"; I wanted to know if they could make it rain.

We drove past groups of huts scattered between large fields sectioned off with brush fences. The sun was already sweltering, the sky cloudless. From out of the bush, a man appeared carrying a walking stick and a plastic briefcase. It was the man whom Elias had contacted because he knew where to find the rainmaker.

"The spirit came to her this morning while she was walking to another village to visit," Mr. Ncube explained. "Now she is in a trance at someone else's home, but she said to come. Mama Mbyata is waiting for you."

We took a little-used red dirt track. As we bumped past clusters of thatch-roofed huts, some with a single solar panel, Elias said quite a few people have the ability to make it rain. He told me that his father had three wives, and one of them had the rainmaking spirit. "She'd fall down hard on the ground when it came into her, but she never got hurt. The spirit comes to them when they are children, maybe ten or eleven. They want to be by themselves a lot. Sometimes it's inherited and can skip a generation, but I didn't get it!" he said, smiling.

We stopped while Mr. Ncube asked directions from someone walking the road, then started off again. He said, "The spirit comes up automatically, and the elders know what to do. They recognize the signs of a trance coming on and arrange for the drummers and prepare something to drink."

Finally, we pulled into the village and walked to the mud hut where we'd been told we'd find her. A row of tall drums leaned against the wall and four women sitting outside told us to enter. In the dark I made out two bodies wrapped in blankets on the ground. A woman leaned against the central pillar. She told us where to sit. I looked around, confused. Were these dead bodies on the floor? Was this where the rainmaker would be?

We sat in silence for a long time. Then something moved. One hand came out from under the blanket, then the other, and suddenly a woman sat bolt upright. Her eyes were closed, and her body shook with spasms. Hands trembling, chest heaving, she made strange sounds: squeaks, moans, huffing. "She's the rainmaker," Elias whispered.

With her back ramrod straight, she began singing. Her eyes were open but unfocused. Mr. Ncube sat on a three-legged stool at her feet. She mumbled something to him, and he turned to me: "She says someone from far away has come in and she welcomes you."

A young woman was summoned by Mama Mbyata and given whispered instructions. The assistant rummaged around in the blankets and pulled out a hat with spike-like protuberances, a blue visor, and a cloth hanging down the back. She lifted the hat with reverence and solemnly placed it on the rainmaker's head like a crown. Mama Mbyata hummed and chanted; then a terrible shaking took over, and just as abruptly, stopped.

A gourd was cut in half and filled with water and the assistant handed it to her. She wet a wildebeest's tail in the water, then drew the tail under her nose, inhaling. She chanted again, stopped, and asked for her rattles. Another search ensued, and one by one, five or six gourd rattles were found in the blankets' folds and put into her hands. She shook one and cried out: "No, no, no!" and threw it down. She tried another and another until she found the right one.

The door opened and four women squeezed into the tiny hut. They sat in the dirt and began singing in harmony while the rainmaker shook a large rattle to one rhythm and a smaller rattle to a different beat, all the time uttering staccato moans. "They are bringing up the spirit," Elias whispered. Then the bundle at my feet moved and a bare-chested woman pushed blankets away. She sat up, her shoulders shaking as if from a fever.

The gourd rattles' syncopated rhythms grew louder. Mama Mbayata thrust her chest forward; her face was blank. She mumbled an instruction: "Tell the villagers to bless their garden seeds," she said, then lay flat and motionless.

No one moved for a long time. When the rainmaker sat up again, she said she'd had a dream: that the children would get sick and have terrible headaches as if their heads had split open and they'd foam at the mouth. The elders were to take them to the borehole and tie sedge plants to the children's hands.

We had already been in the hut for three hours when Elias leaned forward and spoke to the rainmaker. He told her I'd been hit by lightning and certain powers had revealed themselves to me and I wanted to know how to use them.

She asked me to sit facing her. I kneeled and she held my hands. She said there were obstructions that had to be removed first. She drew the wildebeest tail through the water in the gourd and after, pulled my hands along the length of the wet tail, then held it to her nose. I was pushed closer and she wiped the tail across my forehead, then hers. After, she rubbed her forehead hard against mine—back and forth, back and forth.

I was wearing a loose cotton shirt and she wetted my chest with the tail, leaned in, and licked my skin. She asked me to wash my hands in a small bowl of water and I did and then she raised the bowl with two hands above her head like a chalice and drank down the dirty water. She said that to remove the obstruction, I must wash myself in the Zambezi River and make amends with my parents. Mr. Ncube assured her that he would see that I did. After, we sat in silence. Finally Mr. Ncube leaned forward and said, "It's over now. You can thank Mama Mbayata and go out the door."

. . .

Out in the bright air of the village, I looked back: the rainmaker was lying on the ground, eyes open but unseeing. One of the helpers had crouched by her and was wiping sweat from her face and chest. Elias, Mr. Ncube, and I kept walking as the dry shushing of rattles rose up with the harmonies of women singing, and Mama Mbayata's low animal groans merged with the sound of men beating drums in a tight circle around her as the whole of Zimbabwe waited for rain.

*

The next afternoon, Allan and I drove out across the ranch in his old jeep, looking for animals. At a small river we stopped and walked through water to see what game had come to drink there, frightening a kudu bull that ran into the brush, blending in so well, he was barely visible. A family of warthogs crossed our path, tails high, short legs shaking a shrub and sending its winged seeds flying.

Upstream we saw three storks and a large bird called a korhaan. We drove on, flattening a track through thatching grass that opened out onto a vlei. Beyond was a wetland where flocks of birds lived. The patch of sharp-edged reeds in standing water was thick with frogs and pythons.

At the twenty-acre dam, Cape buffalo drank, and white-faced whistling ducks careened by. There were black and white ibis, Egyptian geese, black-winged stilts, Hottentot teal, and finally, a huge crocodile, the green tops of its eyes slowly moving by—Allan smiled at the sight. A high, reedy whistle of ducks filled the air.

At the far end of the dam we turned west toward an encampment of settlers who had been given land by Mugabe. "Only two percent of these people really wanted land or knew what to do with it, how to live out here with all the game. The rest just wanted food and jobs, but this was engineered by Mugabe to stay in power," Allan explained.

"Ironically, many of the farmers being kicked off their land had bought those farms after the war when it was ascertained that no one else wanted them. There are tremendous injustices going on. The worst of them is the eviction and killing of black commercial farmers, like what happened to Billy. It's less about race and colonialism than about tyrannical leaders asserting power."

Two young men were digging a well by hand. A circular garden plot was set out with a clump of banana trees at the center. Women were planting an inner circle of herbs and an outer one of vegetables. Allan greeted the headman. They spoke Shona, discussing the weather, rain, and animals as the women showed me the indigenous maize seeds they were planting.

We were headed back when a wild-eyed young man ran toward us. He had Rasta hair, wore a red ZANU-PF T-shirt, and was carrying a gun. I grabbed the seat, ready to duck, but as he got closer, a smile broke across his face. He and Allan shook hands and set a date for the next meeting about the anti-poaching unit. Allan asked him to spread the word and he said he would. "Already the men are making an outdoor meeting place, sweeping the dirt and placing logs in a circle around a fire pit," the young man said. We waved goodbye. Allan looked at me. "Now you see why it's important to make allies out of enemies. It still works! We would have known if it hadn't—he would have killed us!" he said, laughing.

Every afternoon Allan and I returned to one of the water holes to see what animals were there. It was the time of day when he was happiest. Seeing the animals come to drink summoned up memories of his old life living in the bush away from the agonies of war and tyranny. Armed with binoculars, we sat in the jeep waiting. The croc was half in, half out of the water, and a black ibis jetted upward on thermals. A group of shaggy-maned waterbuck wandered up a rocky knob, their white butts ringed with black like moving targets. A leopard lay on the branch of a tree.

"Some days I'd like to just ride off into the bush on my horse and not be seen again. But then people would think I was just an old crank. I want to be out helping to make it a better world, but it means I have to live a certain way—I have to travel and give talks and try to make things happen—and more and more, as I enter my eighties, I'm growing tired of it."

*

At the end of my time in Zimbabwe a group of us from the Centre went to the Zambesi for the day. We boarded an old boat and passed through the chop of currents. A hippo came up by the side of the boat, its jaw opened wide. The other side of the river was mixed forest and tall palm trees. A hammerkop's nest was wedged in the crotch of a tree, and the bird fed in the shallows above dozing hippos. A monkey swung in and out of view. A game scout's bright laundry was hanging to dry in a baobab, and nearby, a cormorant walked on submerged roots as if walking on water.

The full moon was bloodred and bobbed in heat waves like a severed head, then went white as it rose. "White as the nuts of an ivory palm," Allan said. I had been in Zimbabwe for a month and wasn't sure when I would be back or, if I returned, whether I would have such idyllic afternoons with Allan to look at animals.

. . .

At the end of the "cruise," Allan and I stayed behind to bathe in the river. I'd told him what the rainmaker had said: that I had to cleanse myself in the Zambesi. "It's important you do as she says," he admonished. Standing in the shallows as the river streamed by, Allan splashed playfully, then began pouring handful after handful of water over my head and back. After, refreshed and freed from all obstructions, we left the great river, with its yawning hippos, and very slowly drove the long red road back to camp.

15.

We met in the Seattle airport in the winter of 2004 at dawn. I had just flown in from Fairbanks, Alaska, where I'd given a reading and spent a long evening with friends. A bit hungover, I had trouble finding the C gates and asked a United agent, who pointed the way. When the man behind me asked her the same question, she said, "Just follow that woman in the coat." I was wearing a long Pendleton blanket-coat. A few seconds later a handsome man in cowboy boots was walking beside me, smiling. I felt slightly annoyed. "What?" I barked. I had a bad headache. He said playfully: "She told me to follow you."

At the C gates we found that both our planes had been delayed, and he suggested we go to the bar. We sat at a high table. "Hard night?" he asked. I nodded. "Yea, me too," he said, and introduced himself as Rifat Latifi, a trauma surgeon. He had the broad face and dark wide eyes of the Albanian Kosovars I would meet months later. He ordered two beers and we drank them down. It was eight in the morning.

I asked about his life, about Kosovo. "Have you ever been to the Balkans?" No. "It's been five hundred years of war and genocide. Now Kosovars are rebuilding from the war that ended in 1999. I was born in a mountain village to peasant parents. My father did not go beyond four years of school. My mother had no schooling at all. They had a subsistence farm—a few chickens, cows, and sheep, and no car.

"You may ask how a boy from a peasant village gets to Yale and graduates. I'll tell you: It's by horse cart, by walking, by leaving home never really to go back again. When I was fourteen, my best friend and I decided we had to go to Pristina and enroll in school. He wanted to be a poet; I wanted to be a doctor. We set out to become those things and we've succeeded.

"After high school we continued on to university; then I went to med school. When I saw a notice for a residency at an American hospital in Houston, I applied and was accepted. After finishing up there, I tried an orthopedic residency at the Cleveland Clinic, but I was so bored!" he said, laughing. "One night, at a party, my life changed. I met the head of Yale medical school who, after a single conversation, said: 'You are coming to Yale to do a residency in surgery. I promise, you won't be bored.' " I went immediately.

"That was a long time ago. Now I teach and perform trauma surgery at the University of Arizona Medical Center in Tucson. I love it. Things go fast and I can save lives. But my deeper interest is my telemedicine project. I want to bring virtual medicine to post-conflict, developing nations around the world." He paused, then said: "My mission is to save the forgotten people of the world, because I was one of them."

Maybe it was the hangover, maybe it was the brilliant, fast-paced talk and the cool fury in his voice and eyes, but tears rolled down my cheeks. "I've spent a lot of time living in subsistence villages," I told him. "I understand."

My plane was being called and when I stood to go, he said: "You should come to Kosovo and see for yourself," and suggested we meet there in the spring. Not quite knowing why or what I was getting into, I agreed. "I have meetings there in May," he said. "I'll be with my family. Since the war, they all live in Pristina. They'll like you. You can stay at the hotel and meet anyone you want in Kosovo—the prime minister, the poet laureate, a surgeon, a KLA general. . . . We'll go to my village and I'll show you what happens when there's been genocide and years of war."

In May I flew to Vienna and the next day, boarded an early-morning Austrian Airlines flight to Pristina. Rifat was late getting to the gate, and while waiting, I wondered if I'd been duped. At the last moment he appeared, brash and charming in Lucchese cowboy boots and a bright red polo shirt with the words "Wildcat Surgery" embroidered on the pocket. I could see in his slightly manic behavior that coming home brewed up complicated emotions.

From the plane, Kosovo was a beautiful green bowl surrounded by snowcapped mountains. As we landed, I saw young boys driving cows into fields of waving grass. "It's not how it looks," Rifat said. "It's a bowl of blood." I'd been reading Tim Judah's book *Kosovo: War and Revenge*. "It's not really war," Judah wrote. "It's total destruction. That was the way [ethnic] cleansing happened."

Rifat's nephew, Flamur, met us at the airport and drove us into town. It wasn't hot, but Rifat mopped his brow and lowered the window. "This is the place where a war was left unfinished, where people who have been tortured are still forced to live under the rule of their oppressors."

We passed roadside graves and men scavenging bits of metal to take to the scrapyard for cash, men handing up bricks to patch the remains of bullet-scarred walls. The war began when Slobodan Milošević revoked the autonomy that Tito had granted to Kosovo in 1974, then instituted a siege of ethnic cleansing to remove Albanian Kosovars from their native land. Ninety percent of the population had their jobs taken away. Schools were segregated, the Albanian language was banned, and by 1998, full-on ethnic cleansing was taking place. "They shot the boy of one of my friends," Flamur said. "He had been outside playing. When the parents came out to rescue him, the Serbs grabbed them and forced them to watch their son die." Men and boys were separated from the women, lined up, and shot. Identity papers were confiscated from 1.3 million Kosovars, and of those, 860,000 were expelled from the country.

Flamur continued: "Masked men—Serbs—came to the doors and told people to leave or they'd be killed. They were herded onto trains. It was like World War II. They thought they were going to the gas chambers. Some of the old people were crushed to death. At the last station they had to get out and walk into Macedonia. Some elders didn't make it."

Those not deported were internally "displaced," a euphemism for being tortured and/or killed. Half of all houses in Kosovo were destroyed. The Serbs forced two hundred thousand men into the forest as winter approached and torched their villages. Some were jailed in Serbia. There are thousands still missing. Some are Rifat's family. "When they tried to do away with the Kosovars' guerrilla army—the KLA—they failed."

Rifat told me how, in March, during one of the worst battles in the Serbian war against the Albanian Kosovars, he wanted to leave Yale. "I had only a few months to finish my residency. I went to the dean's office and begged. I told him that I was two months away from graduating. I was already practicing medicine. I had to go home and help the people who needed help." But the dean refused. He said, "If you don't finish, you won't graduate, and you will never be a doctor."

Speeding toward Pristina, Rifat looked suddenly tired. He took a deep breath. "So I stayed and did everything I could to get my parents and brothers and sisters and their families out. I sent money. I hired a driver to take them across the border. There was no way I could know if the driver would really deliver them there, or if he was working for the Serbs. I didn't know for weeks if they were alive or not. They'd had to move twice, and all that time I didn't hear from them. But in June, my phone rang. It was my father. They had made it to Tirana, Albania. The war had just ended thanks to the NATO bombs. I'd just gotten my degree, so I caught the next plane and joined them on the long march back to Kosovo." He looked away, then back. "I was there, but I didn't belong. I had no right to be among those who had suffered so much. But I was happy my family had survived, happy to be with them."

In the days after our arrival, Rifat showed me around Pristina. We walked to the hotel on Bill Clinton Boulevard, so named because of the NATO planes he had sent. "After the war there was no life here at all," Rifat said. "It was hard to find any house standing. The animals had been killed and thousands of people had been deported. There was the smell of death everywhere. But now it is becoming lively again."

We sat in the capacious hotel lounge and drank espresso, but when a Serbian family walked in, the people at the bar stood abruptly and left, unable to bear the presence of the enemy. Rifat was undaunted. I asked if there was still violence in Pristina. He shrugged and seemed agitated: "Kosovo never belonged to Serbia. Kosovars are descended from Illyrians—ancient Greeks; the Serbs are Slavic. But there's no reason we can't all be neighbors. We have to keep asking how to create a viable peace, how to live with our enemies, how to set it up for those in the next generation who have no direct memory of the war," he said quietly.

Matthias Reinecke joined us. A blond-haired, blue-eyed Austrian in his forties, I'd seen him at the airport in Vienna reading *The Enigma of Arrival* by V. S. Naipaul. As part of the European Reconstruction Commission, he helped Rifat get the first EU grants for his telemedicine project.

The three of us walked to a building with a sign that read TELEMEDICINE CENTER OF KOSOVA. INTERNATIONAL VIRTUAL E HOSPITAL NETWORK: CONNECTING THE WORLD. The interior was modern and sparse. Rifat walked energetically down a long pink rug into a room full of computers.

"A virtual hospital functions like an octopus," he explained, hardly able to contain his enthusiasm. "It's a central command center of on-call doctors and has many virtual arms going out into the countryside, so when there's a medical emergency, a local nurse or doctor can talk them through whatever procedures are necessary, from bandaging a wound to performing surgery."

He pointed to a map of the world pinned to the wall. "We are trying to get this going all over the world, and we are starting right here in Kosovo." Matthias continued: "We need to raise the standards of hygiene. Infant mortality is one of the highest in Europe. Yes, you forget that we are part of the European continent. The Serbs would rather not have us here. So if you need heart surgery, cancer treatment, or have an infectious disease, you have to be sent elsewhere. It costs a million euros. Nobody has that kind of money here."

We walked next door to Pristina's main hospital. Rifat said I should see what third-world medicine looks like. A rooster strolled into the emergency room. The walls were crumbling, the floors dirty. Blood was splattered everywhere. Rifat showed me where women came to have babies. "See, there's no place to wash your hands here!" he said incredulously. There were no sheets on the beds. Patients had to supply their own. Their families slept on the floor.

When we emerged from the hospital it was pouring rain, and we quickly found a small restaurant. A beautiful young woman came to the table. "Today we are having chicken in wine sauce, roast potatoes, salad, sweet peppers, fresh bread, and wine. Okay?" A young boy came in selling chewing gum. Rifat bought it all. "I was poor too when I was a kid," he said, "so I always give them money."

A line of UN trucks filed by. "We've changed pashas—occupiers!" Rifat said, laughing. "First it was the Turks, then the Serbs; now we are occupied by the UN." More laughter. "We lost all these lives to get a half-assed liberation? Do they think we're not smart enough to run things?" Laughter. Then the tone turned serious. Rifat said, "This town is full of collaborators. It's like all wars—people are only out for themselves. We've had to fight hard for our country so many times, and yet we still haven't won."

At the family apartment that afternoon, I met Rifat's mother, "Nona," and his sister, brother, and father. Tea was served in tiny tulip-shaped glasses. His parents were small and withered, chair-bound and inquisitive, and they smiled adoringly at their famous son. They still dressed like peasants: his father wore a traditional white fez and his mother wore billowing pantaloons. "I bought jeans and cowboy boots for them when they came to visit me in America, but they wouldn't wear them!" Rifat exclaimed in both English and Albanian. They laughed and explained that he'd tried to get them to move there with him, but they refused. Even though city life in Pristina did not agree with them, they had survived the war and Kosovo was home.

Nona's twinkling eyes glistened. She took my hand and asked me to sit close. "Do you want to know how Rifat came to be?" she asked. I nodded. "It was a warm autumn day. We were harvesting plums to make raki—that's our national liquor. The year before I'd lost a child, a boy, so I went to a woman I trust who can see the future and asked her what would become of me. I thought I'd been bad and that's why the boy had died. It was this same time of year when the plums were piled up, and the hay was forked into stacks to dry, and the squashes were big and orange like moons all over the ground and that woman said, 'Don't worry, you'll have another son and he'll be so powerful and smart, he will amaze you. You can't understand how famous he'll be.' I thanked her and walked home through the old man's field where the hay was cut down. I crossed the stream, throwing a few rocks into the water for fun, then up through the cows to the house, and we drank raki to celebrate. We didn't drink very often. That night Rifat was conceived and in that same room, he was born."

A tray of candies and sweets and glasses of Turkish coffee were set down by us. Nona continued: "It's said that one who is born with the placenta whole and unbroken will turn out to be special. The room where I gave birth was dark and musty, made of rock daubed with mud, and water dripped all along the north side. We had candles for light. My sister-in-law was there. I was about to give birth. I remember seeing my shadow on the wall. It looked like a moving mountain. There was no doctor. The sheep were already lying down around the house. I could hear chickens pecking and clucking, and a low wind. That's when Rifat came out: a big red egg. I could see him trying to get out. I sat up on one arm and clawed at the placenta, trying to break it open, and my sister-in-law pulled from the other side until it split and we lifted the baby into the air. I reached and felt between his legs, for a scrotum—it was a boy—and gave thanks to Allah. I held him up high and said: 'This is Rifat.' "

*

In the week that followed, there was a flurry of activity and Rifat's cell phone rang incessantly. "He gets more phone calls than the president does," Mathias said. Rifat was trying to establish telemedicine centers around the world. "I'm not doing this for myself," he said. "Certain factions of the government are waiting for me to fail. Then they'll just close the doors. My generation has lost everything. What does it take to get beyond this? I have to try."

Rifat arranged a meeting with his childhood friend Milazim Krasniqi, the poet laureate of Kosovo. We had lunch in a big, high-ceilinged restaurant. He and Rifat had walked to Pristina together as boys to enroll in school. Milazim said, "We were just peasant boys with big ambitions. Both Rifat and I wrote poems. In the fifth grade I wanted to send one of mine to the newspaper, but I didn't have the money to buy a stamp. I pleaded with the postman to take it anyway and he did. We only got mail every three months. Finally, I heard that I'd been published. It was the happiest day of my life. Now I've published nineteen books. PEN International helped me. The government here was going to put me in jail, and PEN arranged to get me out."

Afternoon rain came with thunder and lightning, and everyone in the restaurant became frightened because it sounded like the war. Later, driving a back road, we picked up a hitchhiker. Rifat explained: "I hitchhiked so many times from my village into Pristina. . . . I can't resist." The young man smelled bad and wore tattered clothes. Rifat drove out of the way to drop him off at a house and, at the last moment, reached into the back seat and gave him a sweater.

We were on our way to meet Dr. Fadil Beka, who had been a friend of Rifat's in med school. Handsome and soft-spoken, he had been chief surgeon for the Kosovo Liberation Army during the war. Rifat left me off and said he'd be back later. Fadil and I sat in a small room together. He rubbed his eyes from weariness—he'd been in the OR all night. During the war, he told me, he couldn't remember ever sleeping.

"No one can be a good surgeon unless he's operated at the front during a war. If there hadn't been a war here, I'd have had to look for one! I learned so much. Strangely, it was the best part of my life. If only people could learn to work together as we did then."

He pulled two bottles of cold water from his pack and offered one to me. It was a hot spring day and there was no air conditioning in the hospital. "Out in the field, we'd take over a farmhouse. The kitchen became the operating room, with the bedroom used for recovery. When the Serbs began attacking, we had to move quickly. Once, during the night, I helped move two hundred wounded on stretchers and carried the one I'd been operating on over my back. Sometimes we had to go many kilometers before we could find another house that was safe. Then we'd carry them in, wipe off the kitchen table, and set up again. We never lost anyone that way; we never had to leave anyone behind.

"Another night, when we were being attacked hard and we were moving our patients, we almost ran into an oncoming Serb unit. By chance an old friend of mine who was a KLA general diverted us. There had been an accidental bombing of Albanian soldiers with a hundred wounded and he told us to go immediately to them. We changed direction and went to the valley where we were told we would find them and set up an operating room there. In doing so, we saved ourselves. That was the irony of friendly fire.

"But another time, while operating at a little farmhouse, we realized that the Serbs had surrounded us on three sides and there was nothing we could do, no way to escape. We were going to be killed. I had to finish the surgery, so I told someone to get on the radio phone and call the NATO forces. Believe it or not, they came immediately and flew over with bombers. The Serbian forces retreated, and we were saved."

Back at the hotel bar I waited for Rifat. He'd had a meeting with his old friend Hashim Thaçi, who had gone from war hero to suspected war criminal, and later became president of Kosovo. Post-conflict convictions were complicated, Rifat told me. Thaçi had been powerful in the KLA during the war, procuring arms and money from Albania and the diaspora. Later he helped with the peace negotiations but was sentenced to ten years in prison for war crimes. "War is one thing, but what happens to people after a war is something different: revenge and justice don't go hand in hand. After genocide, how can the outcomes not be tainted?"

Rifat ordered two glasses of wine and put a small chapbook into my hands. It was a book of his poems in Albanian. "For you," he said, then informed me that we were to be on an all-night call-in radio show starting at midnight. "Since the war no one can sleep, so it's become the most popular show in town."

I thumbed through the chapbook and suggested we translate the poems into English. "We will," he said, and we walked hand in hand to the radio station. On the way we passed an empty restaurant. Rifat said, "Only Serbs were allowed there during the occupation. Before the NATO bombing, they'd be in there getting drunk and dancing while everyone around suffered. Now they don't dare go there anymore."

The rain stopped and the streets were full of young people going to bars. Upstairs in the studio, the radio host was lithe, beautiful, sensual. While the mic was being adjusted, Rifat whispered playfully in my ear, "Why are we writers? There must be something wrong with us. Something missing from our childhood!"

The host stretched her long legs and gave Rifat a seductive look, then read his poems with an actor's lilt. The calls poured in. Could doctors write poetry? "Yes, of course," Rifat said. Some calls were about telemedicine, some asked questions about life in America, some were about my cowboying life. But many were medical inquiries that Rifat patiently answered.

The host put her hand on his as he talked. The curtain behind them was scarlet and orange, like a wall of flesh. At the end of the show we drank raki and ate a middle-of-the-night dinner. Rifat was in high spirits—rakish and cheerful—but the poem he had written for me was dark. "You're on your way back to Greenland, aren't you?" he asked, and I nodded. Then he read:

> Move above the tree line
> Run away from yourself
> Go where the dark days give way to black nights
> If you go far enough you may heal your ailing
>   heart
> Do not look back
> The mountain of tears is coming down . . .

*

The next morning, I woke to someone hosing down the streets. Rifat burst into my room and slammed the window shut. I had been listening to the imam sing. Rifat said, "We're not religious here. Women don't wear headscarves. They function everywhere—in politics and the arts. The imam is not important. No one cares."

For the rest of the day his mood was somber and volatile. We took off driving. We were on the way to Drenica, where the Kosovo Liberation Army had been strongest and where Rifat was born. His family followed us in another car. By the side of the highway red poppies and wild roses grew between roadside graves.

"It's easy to die, but it's hard to live," Rifat said as we made our way into the countryside. He was driving too fast and a CD of Albanian music was blaring. "Too many damned graves," he blurted out. We passed a young boy on a draft horse moving cattle up a green slope. Farther on, he stopped at the grave of a friend, bowed his head, then wordlessly returned to the car. In the mountains we were stopped at a checkpoint and motioned on.

Rifat laughed. "During the war at that border everyone got stopped. They were pulled from their cars and tortured, because the valley beyond was famous for the battles of the KLA, so anyone going that way was suspect." The valley appeared before us. Ireland-green, it remained unplanted in an area that had once been famous for wheat and corn. "That's because it's pocked with landmines," Rifat said. Then he flashed a wide smile: "But look how beautiful it is!"

The massacre at Drenica was a grotesque expression of ethnic hatred. Men and boys who had been hiding in a nearby forest were lined up and shot; a pregnant woman's body was slit open, the unborn baby flung aside; another's skull was smashed, and the brains were laid inside her body cavity.

Rifat slowed the car. "Drenica was destroyed almost to the ground. So were the people who lived here. I had a classmate from university who was the doctor here. They cut off his left arm in front of his wife and patients. He'd delivered at least three hundred children. They're probably all dead now."

We drove through valley after valley past narrow, pointed haystacks like something out of a Van Gogh painting, but these were framed by gutted villages and burned-out tractors and cars. "The Serbs came to a factory in one of the towns and threw all the workers into the furnace. In another place they collected eighty-four men and executed them. My uncle was one of them. It was like Rwanda here."

Cows grazed between graves. Children filed into a half-wrecked schoolhouse, still in use. We stopped in the village of Perkaz to visit the bullet-ridden house of Adem Jashari, a KLA general who was killed along with fifty-seven members of his extended family. Only two survivors remained.

Three village boys wandered down a winding track tending milk cows. "There are still thousands missing. They found a truck near Belgrade full of bodies, and mass graves next to a military airport there. We won't have peace until we've counted our dead."

Rain began. We drove a bumpy road to a hilltop cemetery where most of the KLA commanders and generals were buried. The view was stunning. I remembered what Hava, Rifat's sister, had said about Kosovo: "The mountains are carrying snow on their peaks, but the land below is carrying the dead."

Rifat had to leave the next day, so his nephew, Flamur, became my translator and driver. There was one more person I wanted to meet: the head psychiatrist of Kosovo, who was an old friend of Rifat's. In the village of Klena we drove down a long lane, then entered an almost baronial gate that swung open onto a farmyard. Anton greeted us. He had been in medical school with Rifat and has been overseeing the mental health of the entire country since the war.

Anton's wife came in with platters of food and glasses of tea. "Rifat and I were together day and night in school. We didn't have other friends. We studied together, ate together. He was so smart and quick. He was writing for the newspaper. He made money for tuition that way and even saved enough to send home. One night I was teaching him about the structure of the larynx, and as soon as I finished the explanation, he repeated it back to me word for word as if he'd recorded it. He has a brilliant mind. We both wanted to be good doctors."

Anton continued: "During the war, there were soldiers and tanks all over this area. One day the Serbs came into this courtyard demanding food. My father went to the gate and politely told them he had no food to spare. I don't know why, but they let him live. They'd put me on a death list, so I had to hide out in the village and sneak home at night. For years we hid the women, children, and my brothers, so it was only my father the Serbs saw. We made it look like he was on the farm by himself.

"Now it's my job to take care of the people who have been harmed by the war. There are many of them. I wasn't well myself for a time either. I had only nightmares and jumpiness. We here on this farm are better now, but we still keep the gates closed."

Back at the hotel that night I heard sirens and smoke wafted through the window. More sirens, then gunfire. I rolled off the bed and lay on the floor, wondering what to do. Rifat's sister, Hava, called. "Keep your head low. I'll come get you when it's safe."

A few hours later she arrived wearing a uniform in an armored military Land Rover with a driver and an aide. The KLA had transformed itself into the Kosovo Protection Corps, a peacemaking group, but her barely contained rage and renewed nationalism were clearly simmering. Its members were almost all former soldiers. "I wish I had been able to fight during the war. It's my one regret," she said as we roared away from the center of town. "Last night one was killed and seven wounded."

She told me that many of the ex-soldiers had remained in training and were re-arming, getting ready for independence, just in case it turned violent. "If independence isn't granted, there may be a coup," she said. They were buying Kalashnikovs for five bucks apiece, but things were quiet for the moment. We drove around the outskirts of the city half the night. On our way back, as we were waved through a checkpoint, I saw her tense up. "Genocide does not leave you alone. It is always in our foreheads," she said, tapping the place between her eyes, then looked straight ahead. "We will always have guns."

The following week Rifat returned to Kosovo looking exhausted. He'd been up all night, operating, then jumped on a red-eye from Phoenix to JFK to Vienna, and finally Pristina. At the airport he said. "I don't know who I am. Kosovar or American? Can I be both? Tell me."

His phone rang. There was an emergency. A young computer technician at the telemedicine office was having a heart attack and Rifat was raced to the hospital. There was no coronary unit, no heart-lung machine, none of the necessary equipment for open heart surgery. There was very little Rifat could do. Helpless and despairing, he finally leaned down and held his friend. The young man died in his arms.

At the end of the week Rifat's family drove back through Drenica, up to his home village of Klondernice. We had to park the cars and walk the rest of the way. There had never been any real roads.

It rained; then the clouds lifted. The path twisted through foothills. Near the top were the footings of Rifat's family farm. Trees and vines had grown over crumbling walls. "This was my bedroom," he said. "This was the kitchen." The walls were pocked with bullet holes and gaps where grenades had been thrown. "They made sure it all burned down."

Nona led me through what had been the farmyard and held my hand tight. "The day we left I went out to feed everything. I gave water and grain to the chickens. I pushed all the hay down for the cows. For the last time I looked into Rifat's room. It was lined with books and pictures, and all his poems collected in big notebooks. When it was time to go, I put whatever food we had left in a place where the Serbs wouldn't find it, in case we came back.

"Leaving was harder than staying and dying. It was like leaving a child behind. That's when I knew I couldn't do it. I told the others to go on without me. Then Hava came over and led me to the waiting taxi. She said, 'Look, we're all together. We're all here and we're alive.' I got in the car and as we started off, I turned and looked behind at all that we had worked so hard to build up. Then I saw the Serbs come down the hill to the farm, shoot the animals, and set our buildings on fire." She stopped and grabbed my hand harder: "Rifat's room with all those books and poems—it burned the brightest."

Later that evening in Pristina I received a call that my house in California with all my books and manuscripts might have burned to the ground. Rifat looked at me tenderly. "Our poems are burning, and our houses," he said. "We will never find a home. . . ." He pulled a sheaf of paper from his pocket and read the poem he had written for me:

> Night has become my bodyguard
> Death leads me to the horizon without light
> Someone is always dying

UNSOLACED

At the grave's door I stand guard with
My scalpel and gloves but
No mask
The injured of my town are waiting
In the hall
Last night I was on call
But no one died
Not even the beautiful woman stabbed
By a long knife
This morning she was laughing
I saw her with my own eyes, I swear I saw her . . .
At the door of the hospital
The wounded of the night
their broken laughter
disappears
Night is the bodyguard
Of my unshed tears.

16.

On my return to California, the burned-over road into Hollister Ranch looked snow covered. White flew as I made my way up to the house, astonished that it was still standing. Inside everything was covered with ash. I didn't bother to sweep up. Instead, I packed, hooked up the gooseneck trailer, loaded my two horses, and with Gabby in the passenger seat, drove to Wyoming.

We humans have made a world where common sense, compassion, and care for one another and the planet has become too much of a rarity. At the Wyoming line I thought of my first sight of the state, of the woman who galloped to the post office at Tie Siding. Now I found myself behind a row of Halliburton oil field trucks. I had entered their filthy empire of ozone alerts, hydrocarbon carcinogens in groundwater, and the usual social ills that come with oil and gas extraction: drugs, prostitution, and domestic violence. Sublette County's Jonah Field and the Pinedale Anticline were drilling year-round. Forty-five hundred new wells were slated to go in soon.

I was perched on some edge even while moving, "hanging ten," looking down on the deep wounds I'd seen. The war dead, genocide, species extinctions, fossil fuel extraction. It was all of a piece, all a form of aggression. Pimping the land and animals.

Profit at the expense of all else. Exclusion and exclusivity. Willful ignorance and denial. As I drove, I tried to register how it felt "not to be."

Once at the cabin I couldn't work, couldn't quite wake up. The pump in my water well had failed; then the solar system went clunk. No lights. No power. I spent much of every day digging into snowbanks, filling every pot and pan with snow to melt on the woodstove for drinking water. Whenever I ventured out, it was to have tea with my neighbor Teddy, a remarkable woman in her late seventies, or to fly along the edge of the mountains with Rita in her blue Cessna. But in the end, I preferred to be alone.

Every day I ventured into a deep valley. The days were nothing. Everything. Even the dead forest had its songs, a numb roar of wind through bare and burned branches where sandhill cranes' end-of-season calls echoed. Nothing is ever wild enough. Everything is wild. Day after day at roughly the same time, I walked the same trail: up and over the moraine above two string lakes, then from one gravel bar to another through thick willows. Where the valley narrowed, a revetment rose straight up—rock walls beneath patches of wild hollyhocks. Near the top, two eight-hundred-foot-high cliffs clasped the narrow river that plunged between them.

Mo lien: Chinese words that mean "grinding-firing," as in the making of bricks, though it also alludes to human creativity. Round trip, if I went all the way to the river crossing, was about ten miles. The valley was half a mile across but perfectly scaled: minute and massive at once, and my rough footsteps on the trail—the grinding—resulted in the kiln firing of a landscape inside my mind.

Every walk was the same. Every walk was a surprise. Each tiny change flooded into the dying present. A day earlier the whole river valley turned gold: water and walls, and the moose and her twins foraging in the willow thicket. The next day the valley was smoke filled, the lake water a dull cerulean, the slopes dun-colored. As I walked a towhee flew beside me, alighted on a stump, sang, and flew off; a wind-battered harebell blossom shook as if trying to ring.

A young elk, in what must have been the first summer on his own, stood by the river pawing water. The splash scared him, and he jumped back, then stepped forward again and extended a hoof. Playing with water was what we all did as youngsters. When the young elk reached too far, he slid down the bank into a deep stream, thrashed, spun sideways, and finally clambered back up. Shaking water off his back, he scared himself again and jumped straight up in place, looked around embarrassed, then settled and started pawing again.

Daily I wandered through the valley's flux of light and movement, its dullness and expressions of exuberance and desire. My forward movement struck the glinting lake: it collapsed in mist. I smelled elk musk and bear. Small surprises dislodged me, and the old alignments shifted into an odd cross between spiritedness and surrender.

Redundancy is liberation. I walked the same path every day. The trail stood for itself and for imposed discipline. Each day, the view ignited, went dull, then bright again. As soon as I turned for home streaks of gold peeled away. There was a lightness to it all, a softness that weighed nothing.

Tracks doubled up like the pages of this book: mine and the coyote's on top of the horse's and the faint tramplings of blue grouse. When my dog Sammy was old, we'd walk this trail down to the creek and sit on gravel. The sight of a merganser gliding by with her duckling clinging to her back delighted him. That was the year I felt the orbital tilt of the earth veer toward darkness, toward unthinkable storms and endless desiccation, and human and animal displacement. I watched Sam have a seizure that wouldn't stop, go into a coma, come out of it, regain his balance and hearing, only to die a week later in pain.

My attempts to understand what Lovelock called "Earth's morbid fever" were fractured and painful. I'd been down on my hands and knees sobbing, I'd gone above tree line, oxygen starved, to hide. Perhaps that's why I kept tramping through that valley: to forget. Perhaps the opposite: to keep stumbling into whatever was there.

Autumn set in, plunking its ravishing blocks of color amid half-dead forests of lodgepole pine. I walked home on the last hike of the year and imagined winter: forty-foot-deep snowdrifts covering the labyrinth of willow, the hidden place where the moose and her calf grazed. A soft sun burned the lake where twelve geese bathed themselves, dipping breast-first into what looked like flames but was water, flapping, stretching their necks. Their honking goose-talk traveled the airwaves in deep musical tones. Two sandhill cranes circled the top of the lake as if trying to tighten it, keep it from draining. At the Narrows—a pinched waist of rock—water from the lower lake pushed into the upper. A trout jumped, catching the last visible bug, and as it did so, the whole valley faded, swallowed from sight.

17.

2004. Panting dogs, jolting sled, sun-bright beveled edges of ice—Jens and I were gliding fast. Beyond was what appeared to be infinite ice and a light-dark horizon that kept expanding as if to give the universe more room. The conditions in Greenland had been growing worse every year. "Since 2000, the ice has been different. Not like before. Not like anything we've ever seen," a hunter said. But my friends Jens, Mamarut, and Gedeon were excited to go hunting, regardless. They'd told me to come early in March when it would be coldest. "Maybe the ice will be good then," they'd said.

I'd returned to Qaanaaq for a four-month stint with David McLain, photographer, Michael Netzer, videographer, and Aleqa Hammond—who later became the first woman prime minister of Greenland—as translator. It was thirty below zero when the last of the dogsleds was hitched up. We were four sleds, fifty-eight dogs, and eight people.

Jens walked in front of his dogs waving his sealskin whip and guided them through the labyrinth of shore-fast ice. As soon as we were out of the maze, the dogs took off and ran fast, exhilarated, and Jens jumped on the sled, laughing. *"Huaquag, huaquag!"* (Go faster!) The sun had just come back. This was the first hunt of the year and I could almost see the dogs smiling as they accelerated. The sled tilted and bumped, sometimes going airborne, and at such moments Jens turned to me with a wordless look of joy.

Indigenous Arctic people moved from the northeast coast of Siberia east across the Bering Land Bridge, through the Canadian archipelago (which we erroneously called the Northwest Passage), and kept going, hopping from Prince William Sound to Pond Inlet, from Ellesmere Island to Peary Land at the top of Greenland. When asked why they didn't go south to make their home in a more temperate climate, they said they didn't know there was a such a place, and even if they had, they wouldn't have known how to live there.

Nittaalaf: it is snowing for no reason. A slight breeze lifted the hair on the sled dogs' backs. Dark patches of frostbite quickly appeared on either side of Jens's eyes and on the tip of his nose, as well as on mine. The leading edges of our fingers and toes went numb as we crossed Ikersuaq, a wide sound separating the town of Qaanaaq from "Steensbyland," where ice-covered mountains were still marked "Unexplored."

We traveled all day in deep cold, stopping twice to make tea, hacking out a piece of multiyear ice to melt for drinking water in a battered pot over a Primus stove. The wind had driven snow into long ridges, and beyond, a curling gray cloud rose from open water. Sun was low as auroral light shot in the four cardinal directions.

"Now we are coming into some good country," Gedeon crooned. But the *hikuliaq*—new ice—was so thin it undulated under our feet like rubber. We set up camp on the edge of the coast, pushing sleds together to make platform beds, and hoisted a filthy canvas tent over a ridgepole. It was cozy inside when the Primus stove was lit. We slept four to a tent. When I laid out my sleeping bag on the far side, the men insisted I sleep in the middle. "It's too cold. You will need us to keep warm."

That night it dropped to forty below. Because it was early spring, the light faded for a few hours and Venus shone.

Late in the evening we walked single file to the ice edge. The hunters' rifles were slung over their shoulders and they held harpoons in their hands. On the way we spotted a breathing hole. It was as big as two fists and we stopped to listen. *"Aaviq,"* Jens whispered. Walrus. We heard them gulping, sloshing, gurgling. Sometimes walrus push straight up through the ice, but these moved on.

At the ice edge a lane of water glistened and blinked in fading sun. Wind-driven ripples caught and dropped the light. An eider duck was perched, unmoving, on a spangled bit of ice as if dazed by the sight of something moving, something as alive as open water.

Gedeon climbed a stranded iceberg to watch for pods of walrus. Three beluga whales swam by, but no walrus. We were silent and after a long vigil, Gedeon clambered down. "Maybe they heard us coming," he said. "But they'll come back. There should be walrus all the way south to Moriusaq. We'll return in the morning."

A gray cloud curled up from the open water. *"Issiktuq,"* Jens said, rubbing his arms. Cold. The moisture in the air made it feel even colder. We walked for an hour back to the tents. Lanterns were lit. At latitude 78° north there were approximately eleven and a half hours of sunlight, increasing by twenty minutes each day. In four weeks, it would be all sun.

We ate a late-night snack of bread and cheese and drank tea. Finally, we slept nose to nose, and despite the snoring, I was grateful for the warmth of my friends.

Before dawn, Mamarut and Gedeon left camp, walked to the ice edge, and harpooned a walrus. Waking up and finding them gone, we followed their tracks on rubbery ice and helped winch the animal out of the water. Gedeon sharpened a long knife, singing as he cut up the thousand-pound beast. "I'm happy the dogs will be getting good food now. They'll have a lot of energy," Mamarut said. Slabs of meat were piled on a blue tarp on the sled. By afternoon, the ice edge had begun to rot and turn gray and it became too dangerous to hunt there, so we packed up and moved on.

We traveled south down the coast. The temperature plummeted. Frostbite masked our faces. When we finally stopped at a tiny hut, I asked Jens how cold he thought it was. He said in English, "Fifty." That meant it was fifty-eight degrees below zero Fahrenheit.

My hands didn't work. I tried to help unload the sleds with my wrists and arms. When Jens noticed, he led me inside, lit the Primus stove, took off my mittens, and rubbed my fingers and arms. In Greenlandic he said, "You must be careful of frostbite. It comes like a ghost and you don't even notice it. Then it becomes very bad."

Two days later we moved on to the tiny village of Moriusaq, where both Jens and Mamarut had grown up. Mamarut's wife lived there and was the schoolteacher. A great-granddaughter of Robert Peary, she spoke a bit of English. When I asked how many kids she had in school, she thought for a moment, then said, "Two."

Jens was quiet. The essence of the living traditions that had been maintained in northern Greenland had thrived because these men chose the old ways. He said, "We like them. We banned snowmobiles and large boats. We wear skins because they are warmer. We travel only by dogsled and helicopter!" he said, grinning. They chose what they wanted of the modern world and forbade the rest.

Our plan was to leave the next day for Saunders Island. The men called it "Walrus El Dorado." But when a hunter came in from that direction, he told us the ice in the channel had broken up and it was no longer possible to get there. Despite a temperature drop, the ice had worsened, battered from beneath by wind waves.

Jens and Mamarut climbed the hill behind the village and glassed the coast in both directions. The ice was now gone from every headland, north and south. We left quickly in the morning. We had made it only partway up the coast when a blizzard set in, and we had to stay in the hut we'd left a few days earlier.

There's always something to do on an expedition. The men mended harnesses, planed the bottoms of their sled runners, and fed the dogs walrus meat. I made copious notes while the men talked and told stories. A great haunch of walrus hung from a hook inside the cabin dripping blood into a plastic bin. "It's like Chinese torture," I said. "The drip drip drip." But they didn't understand the joke.

Greenlandic people are quiet. When Jens spoke, everyone fell silent. He is a "special" person endowed with certain powers. One of those nights in the hut he told an "ancestor story" from the time when animals and people were thought to have a dual existence. With no boundary between them, one was thought to be incomplete without the other. It was said that polar bears could even understand our words.

"When I was little my father saw the track of a polar bear," Jens began. "He followed it and came very close. He lifted his rifle to shoot the bear—he needed new skin clothing—and just then, the bear turned his head: he had the face of a human. The bear was smiling and said, 'Take me, I'm yours.' The dogs were scared and ran away. My father tried but couldn't shoot it. He let the bear go. If a person has special talents, animals will come and ask you to be a helper. You are only asked once by a polar bear, but my father said no to him and the bear ran away."

We spent the next three weeks trying to find ice strong enough to hold us. When the weather cleared, we went up and over the ice cap. Crevasses gaped; we took air over a cornice, then careened down the other side at leg-breaking speed. Dogsleds have no reins or brakes. Jens put soft ropes under the runners to slow us, but it didn't do much good. He used his feet and his knees to slow and turn the sled as we slalomed down a rocky streambed. At the bottom we saw a polar bear track. Perhaps the bear had heard Jens's story and wanted him to follow.

The dogs had not eaten. We traveled up a narrow fjord and saw a drying rack full of meat left behind by the last hunters who had camped there. Jens pointed to it, smiling. "It's the custom to leave meat behind for whoever needs it. It might be for us; it might be for the polar bear." That night the dogs ate two-week-old meat and we made soup and drank tea. Jens looked at his team. "Man and dog go together. It's a good combination. We have affection for each other. We need each other. We belong to each other. Where they go, I go."

During a rest stop, Mamarut spoke about polar bears. "The bear has no weapons. He is himself a weapon and can move on ice or water equally. We love seeing them. We admire them so much. He can walk on very thin ice or lie flat on all four paws and crawl. If he slides into the water, he uses his paw and leg to climb out sideways, and if it happens to us, we do the same."

That night Jens told us a bear story: "I was in the tent when my dogs started barking. I knew it must be a bear and I grabbed my rifle. A polar bear was standing on a little hill of dirt and ice. He beckoned to me to come with him. I knew if I went with him, I would not be able to come back, so I said no. I didn't want to desert my family and the community. This is the modern world and there is no place in it any longer for the person I would have become."

He said that sometimes he knew when things were going to happen. "I feel it, then I tell others, and it happens. I'm afraid of this ability. Sometimes it's vague, but it's there, and it's scary because I don't have control over it. I foresee things, and even if I fight against it, it happens anyway. I knew one other person who was like this. He was Kaliapalu, the elder son of Robert Peary and his Greenlandic 'wife.' He could travel in his mind. He would sit very still, and I could tell he was somewhere else." When I asked Jens if he was able to do this too, very reluctantly he said, "Sometimes . . . yes."

I knew that the next day's travel would be difficult, and I quietly rehearsed the polar bear's lessons of how to get out of the water if I fell through the ice. In the morning Mamarut jumped from ice pan to ice pan, poking it with his iron *tuuk* to make sure it would hold the sleds. When he returned, he and the others talked among themselves using hand gestures in the air to outline the coastline and the difficult passage to Kiatak Island. Finally, a route was decided upon.

The temperature hovered at thirty-five below zero. "The ice is dangerous," Jens said, "and we'll have to move quickly." He walked in front of his dogs through a tumbled wall of shore ice, and as soon as he turned toward the sled to get on, the dogs took off. He made a flying leap; I grabbed his anorak and pulled him on. All around us were moats of open water and pieces of rotting ice—a jigsaw puzzle, whose pieces did not quite meet. Ice pans sank down under the weight of the sled as we passed over them.

We stayed close to the coast at first, then veered out into the open strait. The dogs leapt over patches of rotten ice and their feet sank deep into slush oozing with water. I pointed to their wet tracks and Jens nodded, remembering that we were near the place where we had gone through the ice in 1997. Seven years later the entire Arctic ecosystem was collapsing.

At the north end of Kiatak Island we followed an ice foot, a ledge of ice stuck to the side of the island. Jens shortened and knotted the trace lines, securing them with a walrus ivory stick. We hit down so hard on rough ice I couldn't see for a moment and later found I'd cracked a tooth. Where the ice ledge broke away, we had to jump off a cliff. The men unhitched the dogs and belayed the sleds over the steep side; then, one by one, holding the lines, they went over, pulling their dogs with them. I was one of the last to go. Mamarut was afraid I'd break an ankle if I jumped, so he tied a rope around my chest and lowered me down the face of the cliff like a sack of grain.

Gray ice broke under us. The dogs splashed and scrambled as the men quickly hooked them to the sleds. Jens yelled words I'd never heard before, and the dogs, aware of the danger, lurched forward, jumping across open water. Those scary moments stood for climate uncertainty everywhere and the anguish we felt about the world's refusal to respond.

At last we made it through rough ice to a tiny hut on the west side of Kiatak Island. After a meal of soup and cookies, an apparition materialized: fifteen beautiful white dogs pulling a sled running quickly toward us. A young man stepped off. He was Aviagaq from Siorapaluk. We welcomed him and gave him something to eat. He was fifteen years old and had quit school for the year to see if he could become a hunter. His dogs were good, his skin clothes were perfect. But he was unaware that the sea ice was in decline, and Jens had to tell him that to become a full-time hunter would no longer be possible.

I followed Jens and Mamarut up a steep hill to search for a route to Siorapaluk. We surveyed the frozen wreckage before us: broken ice and open water was all we could see. We talked about the year 1998, when spring failed to come and there was solid ice all the way to Ellesmere Island. "Maybe now we are coming to a time when it will be summer all year long," Mamarut said, and we didn't doubt him. The early spring sea ice had already come apart and we would be lucky to get back to Qaanaaq at all.

Jens glassed the ice out in the middle of the strait. "Too far," he mumbled in English, meaning that it was too deep out there for walrus. They needed a "shelf" on which to get food, and drift ice to sleep on. Jens let the binoculars swing from his neck. "We had everything here. Our whole culture was intact: language and hunting traditions. We kept most of the old ways because they work better. And we took from the modern what we needed. We could live any way we wanted, and we chose this. We like it this way, we can feed our families, and no one tells us how to live. Now I don't know what's going to happen. If the ice doesn't come back, it will be a disaster."

The following day we set out for Siorapaluk, whose population had dropped from seventy to twenty-five since the last time I'd been there. The ice was good enough to stop for tea, and I saw Gedeon banging his ankles against his sled. "I can't feel my feet!" he said, laughing. Mamarut told his younger brother to stop acting foolish. "Springtime is most dangerous of all because as soon as it gets warmer, we forget to be careful and the frostbite comes anyway."

When we arrived in the village, we found there was no place to stay, so we fixed up an abandoned house, swept floors, lit lamps and the Primus, made soup noodles, and slept. The temperature had warmed to minus twenty and for the first time it felt balmy. Happiness returned. The dogs howled in a wave starting up on the hill at Ikuo Oshima's house all the way down to the water.

For a moment we forgot how bad things were until Eva, the Norwegian schoolteacher, visited. "It's now the first time I've seen this northernmost village with no ice," she said. "I've been here twenty-five years. Not enough to eat now. Some hunters had to kill their dogs who were starving. The younger generation I have been teaching will not be able to live here. They will still know how to drive a dog team, hunt, and sew skin clothing, but they will have to learn new skills. Inside themselves they will be living two lives."

Up the hill Paluunuaq Duneq greeted us. A diminutive woman in her seventies, she was wearing handsome sealskin kamiks. Her name means "a polar bear lying flat on thin ice." We'd first met her in Qeqertat, a subsistence village at the head of Inglefield Fjord, when we were hunting narwhal. She poured coffee and lit a small candle on the table. "I was born in Qeqertat inside the fjord. It was a place with a lot of people then. Many families lived there because the hunting was good. Our houses were made of turf and stone and heated with narwhal and seal oil lamps. It was warm and nice. We built the walls thick. We didn't freeze!" she said delightedly.

"The animals seemed to be fatter in those days. They didn't have to travel as much. They weren't scared of noises. After hunting we'd dry the skins. In spring we lived in sealskin tents until late September, then built a new turf house for the winter lined with sealskins. Life was nice then. Everyone had a task, even the children. There was no cloth—everything was made from skins.

"I met my husband there. We had eight children. They nursed until I had no breasts left. The first few babies I had on my own. The others were born with a midwife. I had to sew all the clothes. You can't survive without polar bear pants and anoraks, kamiks, and mittens. It was a lot of sewing for ten of us. I've rubbed the grooves off my fingers," she said, and showed me her hands. "For fun, we had dances. I had a gramophone and we sharpened the needle with my ulu.

"My father told stories of shamans. They were in every village. One story I can't forget. My father was out on the ice when he felt something close and saw a polar bear on his elbows like a human. The bear came closer and my father lost consciousness. Just before, he felt something go into the top of his head, and when he came to, he felt a great pain there. The dogs barked at him when he came back to the tent. They saw someone else inside him. My father became a drum dancer. He was the local shaman. We had a lot of them," she said, and picked up a blue plastic dustbin to use as a drum. Beating it slowly with a stick, she sang.

On our last night in Siorapaluk, Aleqa and I played bingo at the school with the other women, and the next morning the dogs were harnessed and we headed out. It felt warm; the sun was out and despite the bad ice, our moods had lifted. Beyond the mouth of the fjord a mirage held an entire island in the air. We had a dog race: first Gedeon was in the lead, then Mamarut. The dogs got tangled, and the men laughed as they sorted them out.

We hacked off a piece of multiyear ice and melted it for tea. In afternoon light the mirages blossomed. They stacked up one after another against distant islands. One lifted Kiatak on quavering stilts while another mirage made open water flow upward through a cloud. "That's how we know it's spring," Jens said, in his purring, gravelly voice, then called to the dogs to trot down the well-traveled ice trail toward home.

*

Between the end of that ill-fated walrus hunt and the beginning of summer, we stayed in Qaanaaq. At the terminus of one outlet glacier, the spilled rubble was a jumble of ice blocks that looked like dumped furniture. We slid by a low-lying cloud, a humped mountain that had been sanded smooth. Above, the ice sheet was cloaked in white flannel. A raven appeared in a shaft of light, cocking its head to look at us. "Welcome home," it seemed to be saying.

We were greeted by wives and children when we arrived, and though they understood we had no fresh meat, no one complained. We rested for two days. The temperature rose above zero and the nights were bright. In the communal shop, which was outfitted with the latest tools from Denmark, Mamarut and Jens made new kayaks for the coming narwhal hunting season—slightly bigger, to fit their expanding girth—while Aleqa and I visited elders.

As we walked through town, young Aleqatsiaq Peary joined us, asking for a job. But when I told him where we were going next, he said, speaking English: "I don't go out hunting narwhal. It is maybe a little dangerous. No, rock and roll is for me. We're thinking of calling our band the Losers. I'm Robert Peary's great-grandson from when he was with his Greenland 'wife.' I have Greenland eyes. They are beautiful. They get me a lot of girls."

We visited the small museum that had been Knud Rasmussen's home. David Kiviok, the local archaeologist, showed us some of the treasures: the walrus penis bone that was used as an ice scraper, the skin of an auk that was used for a washrag, a seagull hide dish towel, a blubber lamp, and bearded sealskin thimbles.

At the Elder House we visited Minik. "My mother was Arnarulunguaq, who went to Canada with Knud Rasmussen. She had a baby with him and left that child with a couple in Point Hope, Alaska. Then she came home and got pregnant—got me! I was born on Qeqertarsuaq in 1925. That same year my family moved down to Dundas Village and we lived there until we were moved by the Americans, who took our land to build Thule Air Base. My mother died early of a burst appendix, so I was out hunting with my father all the time. We were always on the move. I skipped a lot of school. I had sisters and brothers, but they passed away when they were little. I was the only one who lived.

"My father built my first Qaanaaq-style kayak when I was four. He towed it behind his. Got my first seal when I was eight. There were no motor sounds then, so there were lots of animals everywhere and they were easy to get because they weren't scared. I remember Peter Freuchen. He was running the store in Dundas Village after many trips across the ice to Canada-side with Rasmussen. He spoke Greenlandic perfectly. This is how I remember him: his right foot was just a round thing. He had lost all his toes.

"When I was young, I married my wife and we lived in Savissavik. Very good hunting there. That's where the meteor was found, and people made knives from the iron. I got eight seals a day. We lived in a stone and turf house with narwhal oil lamps. I never liked white man's houses or food. I only like meat. My favorite is walrus meat, narwhal liver, and *mattak*, the fat just under the skin where all the vitamins are.

"We used driftwood to make sledges and kayaks, and covered the kayaks with skins. I used narwhal and walrus ivory for harpoon heads. We had a special stone for sharpening them. Narwhal hunting is the best thing you can experience in your life."

We thanked him for talking to us and on our way out, another elder, Mamarut Simiaq stopped us. He wanted to join the conversation: "Nothing changed at all from our traditions until the 1960s," he said. "We didn't need it, but it changed anyway. Winters were hard. The sun goes down for four months of the year and you can only see when there's a full moon. I like it best when life is so strong it conquers the dark."

On the way back to the guesthouse, mist rose from patches of open water and I saw how those small clouds created a feedback loop. They held in warmth, and beneath them, more and more ice melted. In the distance a polar bear crouched on a piece of floating ice.

Near the guesthouse I stopped to visit Sophie. In her seventies now, she said she had helped the local shaman in the old days. She offered me coffee and cake and lit a small candle. "There are ghosts that circle my house," she said. "They float because they have no legs. Some days everything is confused. My mind too. But other days, I remember everything. Then the ghosts vanish. Even the mirages go away." She stood, picked up a hand drum, and began dancing in place. Abruptly she stopped, put the drum down, and looked blankly at me. Finally she spoke: "Sometimes everything is clear when there is nothing to see."

*

On an early July afternoon, we loaded kayaks and camping gear into Jens's small boat and headed for Inglefield Fjord for a month of hunting narwhals. Gedeon was ahead of us—his boat was faster. "He's young," Jens commented, and made the motion of someone going fast. The others were with him, the women and children and grandchildren. The clouds dropped just as we started out and we couldn't see. Ice floes were in front of us and we moved even more slowly to avoid a collision. A dovekie flew up out of the water, bouncing on the crests of tiny swells. Gulls twisted out of patches of fog that lay behind moving icebergs.

Though it was summer, the nights were still below freezing. As soon as the sun reached the northernmost point of its elliptical orbit, the breeze turned suddenly cold. We had almost nothing to eat because the shelves were empty at the store. The supply ship had been unable to get that far north, so we made do with what we had for breakfast: stale cookies and tea. Passing a gray cliff, there was a sun dog: that meant a storm was on the way. Dark clouds had already gathered in the west and the open water looked like polished coal. Jens had set out a net the night before and caught enough Arctic char to feed the extended family for a few days. As we put-putted along, Jens excitedly pointed: *"Iqqaigaa nanuq!"* It was the exact place where we had helped Gedeon get his polar bear.

Jens is *ilihamahug*—wise—about things, and besides being prescient, he also remembers every moment of every trip we have taken together since meeting in the mid-1990s. The night before I'd dreamed that at a dinner party in Nuuk, none of us wore shirts, just polar bear pants and kamiks. I kept introducing myself with the wrong name, then confessed I didn't really know who I was. Perhaps I had been too long in a strange land that had long ago begun to feel like home.

We cruised alongside what's called "an Inuit picket fence": huge icebergs bright as searchlights. Rays of sun shone through slotted mountains, but even in that long-stemmed light we became lost in dense fog. Gedeon was waiting for us on the other side of the fjord. The fog blinded us, and we worried about slamming into drift ice. I asked Jens where the life jackets were. He laughed and shrugged, held his arms in close and shivered. He laughed again and I laughed too. What better place to die than right here with this extraordinary friend.

Finally, Jens shot his rifle straight up. In response, Gedeon shot his. Echolocation worked: we aimed toward the sound and found him. A storm came in fast, so we headed into the wind, hugging the shore, and traveled in tandem. In the bridge of Jens's boat, a harpoon head swung from a rope like a metronome. The water was gray, the mountains were purple, a scallop of rain splattered across the deck. We passed a chunk of ice that resembled a narwhal's tusk—a form of "tooth" so sensitive it can detect changes in barometric pressure.

Up the hundred-mile-long fjord we went—a ragtag group of men, women, and children. The weather worsened and we made our first camp at a place called Tikerausat in summer snow.

"Qilaluaq! Qilaluaq!" the women yelled. Narwhal coming. I had joined the hunters' wives—Ilaitsuk, Tecummeq, Birthe—and climbed the mountain to glass the water and warn them when a pod of narwhals entered the fjord. The water was choppy and seagulls circled overhead, crying. Lithe as a cat, Gedeon paddled out to a piece of drift ice. With one hand he reached out to steady his kayak against the ice while he waited, bent over almost flat. His other hand was on the harpoon, lying on the deck.

I heard narwhals breathing, gulping, sloshing; then I saw them, black and spotted in the water. They came from behind and surrounded Gedeon. Water roiled, narwhal leapt and dived. All at once Gedeon took off, paddling hard into the turbulence. Paddle in one hand, he hurled the harpoon with the other. When it hit, the breakaway snapped and instantly, Gedeon reached back and pushed the *avateq*—the inflated seal-blubber float that was attached to the line—into the water.

. . .

We camped together for a month, hiking up the mountainside, glassing the fjord, carving up meat, stacking it under a pile of carefully placed rocks to make *kiviok.* The icebergs were our chronometers, but I'd lost track. What day? Which night? Most of the hunts were between one and six in the morning, when it was cool. We tried to sleep during the day. The children played at all times. There was no real bedtime for any of us.

The iceberg that parked itself near camp grew glossy in late sun. Under its turquoise flank, a line of water droplets fell. As always, Mamarut was the jokester. While he was imitating narwhal breathing, the iceberg calved its entire facade, twisted, split in half, and fell backward. "It died," Mamarut said, and feigned a cinematic fall.

August first. Scattered rain, choppy water, hard gusts. We watched an iceberg swim the wrong way, back toward the glacier from which it had calved, as if trying to turn back the clock, trying to stop climate change. "This isn't our usual summer weather," Mamarut said, suddenly serious. "It's usually very calm. The water has to be smooth for the kayaks or else we'll tip over. Jens almost died that way a few years ago. We don't do Eskimo rolls. That's for tourists," he said, and glassed the fjord with his binoculars before letting them fall. He turned to me: "We didn't need to know how to roll. We learned to paddle kayaks in the old days when there was no wind. We don't know how to call this recent weather. It is foreign to us and has no name."

Finally the weather smoothed out and the glassy water was cut apart by the indigo shadows of moving icebergs. The children pulled their driftwood toys with sealskin thongs along the beach. Another narwhal was harpooned. The air turned frigid and a middle-of-the-night wind churned bits of ice onto the beach. "That west wind from Canada-side means winter is near," Jens said quietly.

One of those times on a cold August night after Arctic char soup and rice and a celebratory glass of wine from two bottles I'd brought, we sat lined up in front of the tents in pale sunlight, with parkas and sealskin kamiks on, legs stretched out straight, the way we sit on a dogsled. No one talked. We tipped our faces up to the cool sun and for a moment forgot that soon Arctic ice would be a thing of the past, that the world as we knew it was vanishing.

18.

Knowledge traveled north and south. I'd been jumping between the far north and Africa and kept touching the tips and spikes of climate change as well as those unexpected moments of solace, but some internal churning, a chronic restlessness, kept me returning to the Arctic. My friends in Greenland said I must have been born on a moving dogsled because I keep reappearing there, deeply happy on Jens's sled, blasting and banging over rough ice, but never cold. Icescapes unscrolled before me: the stacked plates of pressure ice, white bears in whiteouts, stranded icebergs, and the black tips of somersaulting ravens' wings. By 2007, our innocence about the effects of global warming on Arctic sea ice was forever lost. Multiyear ice and *hikuliaq*—newly formed sea ice—had almost disappeared.

That year began hot and ended hot. On an expedition backed by National Geographic, I visited indigenous Arctic people around the top of the world to see how they had been affected by climate change. I visited Wales and Shismaref in Arctic Alaska, rode the sleds of nomadic reindeer herders at the edge of the Barents Sea in Russia, revisited villages in Nunavut, and finally returned to Greenland.

So many unimaginable changes had been taking place, and too few had taken the time to understand the chemistry, mechanics, and amplifying feedback loops that kept global heating on the rise. News from what I called the "Ice Desk" was disastrous: the Desk had melted. As albedo decreased, the planet warmed, and more snow and ice disappeared. Earth's forehead was hot from fever.

Was it possible we hadn't fully comprehended that we were in danger—that our species might die off out of carelessness? A research vessel discovered a fountain of methane one thousand meters wide erupting out of torch-like structures escaping from the floor of the Laptev Sea. Gas bubbles were reported along the Atlantic coast from North Carolina to Massachusetts. Oceans were acidifying, and trillions of cubic feet of frozen methane clathrates would soon begin to thaw.

Warming ocean water slid beneath tongues of ice, causing the glaciers of the West Antarctic Ice Sheet to break away. Ash from western wildfires landed on the Greenland ice sheet, darkening it and causing the two-mile-high mountain of ice to become a heat sink.

At the airport in Greenland, the ground crew were in T-shirts on a spring day when the temperature normally should have been zero. On the way to Qaanaaq, I stopped first in Greenland's capital city, Nuuk, to stay with Aleqa Hammond, now a member of Parliament. We went to dinner at one of the minister's houses: Thai food, French wine, and lively, informed conversation. "It's as good as New York here," Alfred, the minister of the environment, said, a twinkle in his eye. Next to me was Laila Fleischer, the first Greenlandic woman to become a helicopter pilot. Across the table a nurse said she was working with Greenlanders who still tested positive for TB—an old scourge that had been dealt with sixty years earlier. "Perhaps the body has a memory for it," she said, "and it has become part of life up here."

Out the window we kept track of an ongoing lunar eclipse and watched the aurora pulse in the north sky. We talked about the hot earth, diminished ice, and the gutless politicians who had done nothing about climate change. Alfred looked dismayed. Filling our glasses with red wine, he said, "We are walking backwards into the future. We need to turn around."

As we flew north the next day, the icebergs' rounded shoulders suggested premature old age. Greenland had already been declared "a rotten ice regime" by Arctic glaciologists. The termini of every glacier were choked with calved ice. The sight of ice trapping ice seemed ironic, as if the glaciers were trying to get it all back, hoarding whatever remained.

It was shocking to fly over places where I had traveled by dogsled—from Uummannaq to Illorsuit, from Moriusaq to Etah—and see almost no ice. The twin satellites known as Grace had been measuring sea ice thickness and revealed that drastic thinning was taking place. Sea ice that was normally ten feet thick in the winter and spring was now only seven or ten inches thick.

A strange mood prevailed in Qaanaaq. Every evening people stood on the hill and looked out to see if the ice was forming. Nights I heard gunshots. The less-able hunters were shooting their sled dogs because they had nothing to feed them. In 1993, the choruses of howling sled dogs in Ilulissat, Uummannaq, Qaanaaq, and Siorapaluk were the all-night operas to which we happily fell asleep. Now the odd silence kept me awake. The only sound I heard was the dry hiss of wind blowing snow off the ice sheet.

I asked Gedeon to take me to Siorapaluk, normally an easy six-hour trip straight up the coast on a route used so frequently it resembled a two-lane highway. This time the dogsled had to climb over an edge of the ice cap to get around a headland of broken ice. The trip took twelve hours.

Siorapaluk was once a bustling village of avid and renowned hunters and their families. Made famous by the French anthropologist Jean Malaurie, it maintained a fluctuating population of fifty to seventy, a small school, a store with a public telephone, and a "skin house," where skins were tanned. Now half its houses were empty, and the school was being closed. On arriving, I walked up the hill to Otto and Pauline Simigaq's house. He was one of the elite hunters, and the skin clothes that Pauline made—especially the mittens and kamiks—were thought to be the best in Greenland.

Otto couldn't sit still while we talked. He kept leaving the room: the subject of sea ice conditions was too painful. "Seven years ago, we could travel on safe ice all winter and spring and hunt animals. We didn't have to worry about food then. Now it's different. There has been no ice for seven months. We always hunt west of Kiatak Island, but the ice doesn't go out that far now. The walruses are still there but we can't get to them."

Pauline handed me the sealskin mittens she had made for me. I inspected the intricate stiches and tried them on. "Beautiful," I said. She seemed uneasy. "We are not so good in our outlook now. The ice is dangerous. Now when Otto goes out, I wonder if I will see him again. Around here it is depression and changing moods. We are becoming like the ice."

That night I slept with the window cracked open, frigid air and sunlight streaming over my face. My dreams were vivid: a volcano spewed blocks of ice instead of lava, and the moats between bits of drift ice filled with small fires. A week earlier, Gedeon had almost perished when out hunting. The ice he was standing on broke away. He cut the dogs loose from the sled.

When a helicopter from Thule Air Base rescued him, they lifted his dogs first, then pulled him up. As he crawled into the helicopter's side door, the ice melted and his sled disappeared.

Now I have a recurring dream about Gedeon. He is carrying a pane of ice by a handle across the sound, laying it down, trying to fill in the open gaps, but the edges of the pane keep melting and the ice floor he stands on dissolves into the sea.

In Siorapaluk I found my old friend Ikuo Oshima at the skin house where he tanned hides for clothing. A Japanese native who came to Greenland in 1972 with the mountaineer Naomi Uemura, Ikuo found a home here, learned Greenlandic, married an Inuit woman, and never left. "I was going to the University of Tokyo in engineering, but the thinking didn't go anywhere," he said. "It was already all figured out and I became bored." Handsome, bright, and fluent in English, he told me that since my last visit his wife's mind had "reversed." He lived alone, hunting with the help of his two grown children, but his optimism remained. "Every year we go north to Etah to hunt, though we get musk oxen instead of walrus. But the meat of both animals is good. Oh, it's so beautiful up there. Yes, I think we shall survive somehow, maybe just on beauty."

19.

Cabin and cosmos, sun and home, and a garden full of radishes and Swiss chard. So much I hadn't had for a long time, yet I missed Jens and the dogs and the feel of sea ice under me; I missed lions roaring and picking thorns from my feet in Africa. In both Africa and Greenland, I'd seen the two root causes of climate change: degraded and desertified earth caused by ineffectual rainfall, and the loss of albedo because of the disappearance of snow and ice.

A Cibecue Apache elder once said, "The purpose of sacred places is to protect the place and to perfect the human mind. Wisdom sits in places." Land instructs us. So do thorns and ice. But we have to be willing to listen and see.

My one-room, off-the-grid Wyoming cabin was built single-handedly by a friend at the northern edge of the Wind River Range. The land is a glacial moraine nicked with kettle ponds and humped with erratics so perfectly placed, they mark the path the glacier took as it passed by eleven thousand years earlier, and though the ice was gone, I could still walk its course. Perhaps that's what connected me to the icescape of Greenland.

Before I ever climbed on a dogsled, my feet knew "the way of ice."

I'd designed the cabin on a cocktail napkin in San Francisco and handed the rough sketch to an architect friend, who had someone print out a plan. When I gave it to Mark Domek, he said simply, "Okay." He cut beetle-killed standing dead trees on the Greys River and peeled, milled, stacked, and chinked the blue-streaked lumber, which he cut square on the inside with corners fitted together perfectly.

Some days the moraine and its ice-carved ponds seemed to sweep all the way into the cabin as if on an incoming tide so that outside and inside began to feel the same. Across the meadow the land bent up into high "parks" of forested slopes where elk grazed in morning sun. Here, I could imagine a wall of ice sliding by and feel its cool wind. Moving glaciers made this place: square-top mountains and thousand-acre meadows, hanging valleys and waterfalls. Ice was the clock that signified time with no before or beyond, only the grinding, melting, firing present. Ice was time's ephemeral source and demise.

So often we miss the whole-fabric aspect of where we live, and our own consciousness embedded within it. We are not interrelated but "intrabranched": one branch wound around another and fused into a single embrace. Our lacelike nervations have overlapping frequencies. It's what the Greenlanders simply call *sila:* consciousness, weather, and the power of nature as one. If nothing else, we are what the physicist Richard Feynman called "scattering amplitudes," wholes within unbounded totalities.

Yet on nights in Wyoming in that hot year of 2007, a dead wind pushed smoke across the withered valley, its torn pastures. Lacking water, sagebrush didn't droop; it went rigid and took on the stunted elegance of bonsai. Much had burned and there was a 90 percent die-off of lodgepole pine, Douglas fir, and whitebark pine. To save our planet we'd have to protect and regenerate billions of acres of grasslands and forests, but so far, there was insufficient initiative for doing so.

Physicists have recently discovered "quantum entanglement," in which paradoxical and counterintuitive measurements of entangled particles result in wave function collapse, or as one scientist put it: "Seeming impossibilities occur, as when a photon that has not been born comes under the influence of one that was already dead." That's how those summer months felt: the boundless ways in which episodic fragments of experience and memory were torn away, sewn into a future that had not yet taken place, then conjoined.

During a chance meeting, the naturalist E. O. Wilson advised me to give up thinking we are doomed. "It's our chance to practice altruism," he said. I looked doubtful, but he continued. "We have to wear suits of armor like World War II soldiers and just keep going. We have to get used to the changes in the landscape and step over the dead bodies. We have to discipline our behavior and not get stuck in tribal and religious restrictions. We have to work altruistically and cooperatively and make a new world."

Altruism? I asked, and he shrugged. We see cooperative behaviors everywhere: ducks performing the broken wing display to lure predators away from the nest; male baboons forming temporary alliances to ward off outsider males. Arctic people sharing food even if it means they will starve.

Are we innately selfish, with only a veneer of social ethics to keep us from incessant nastiness? Or are we hardwired for cooperation and altruism as a necessity for getting along or favoring kin in situations where the survival of many is at stake? Are we designed for anything beyond the functions of the autonomic nervous system? Hasn't it been shown in times of duress that concern for others around us can be eliminated, as during China's Cultural Revolution? But the opposite can be true, as during the natural disasters—hurricanes, floods, and fires—now besieging us as a result of climate change.

Mired as we are in what Wilson calls our "rolling waves of destruction," we persist in the illusion that we are apart from or altogether exempt from the natural order of things. But we are, above all, a deep part of nature—opportunists capable of destruction as well as restoration, of creative problem solving and unquestioned acts of kindness.

Impending doom, pandemic illnesses, ice loss, and possible extinction made me feel untethered. The day I visited the glaciologist Jason Box in Copenhagen, the sun was out, and we sat on a canal lined with small boats and drank hard cider.

"The ice sheet is melting at an accelerated pace," he said. "Two hundred eighty gigatons of ice is lost each year. It's not just surface melt, but the deformation of inner ice. I'm tracking multiple feedbacks and connecting the dots. Beyond surface ice melt and the natural drainpipes—the moulins—there's a drawdown of the inner ice caused by impurities like wildfire ash and industrial soot that darken snow and ice, reducing the albedo effect. It causes melting inside and out. Now cyanobacteria are interacting with black carbon, hooking onto it, allowing it to be drawn into cryoconite melt holes.

"The whole fabric of the ice sheet is coming apart," Jason said. "The subnivean lakes affect the plumbing of the ice sheet. Now 1.5 trillion tons of carbon is being released from melting permafrost into the atmosphere. There will be temperature spikes in summer heat to come. Even a minor decrease in the thinness of ice can double the average yearly rate of ice loss. And for every 1.8-degree Fahrenheit temperature rise, wildfire activity doubles."

Jason reminded me that my "global fox-trot" at the top of the world meant I was skating on disaster: the apron of permafrost that encircles the Arctic had begun to melt and release ancient gases—carbon dioxide and methane—which was accelerating global heat.

The ongoing animal extinctions were dramatic, and a scientist at the University of Aarhaus in Denmark said that entire branches of Earth's evolutionary tree were being chopped off: "If current conservation efforts are not improved, so many animal species will become extinct during the next five decades that nature will need three to five million years to recover."

Efforts are being made to stop extinctions and rewild the planet. At Pleistocene Park in eastern Siberia, the father-and-son team of Sergey and Nikita Zimov in the Sakha Republic realized that if grasslands once stretched from Beringia to the Yukon and Arctic Alaska, it could happen again. They acquired fifty square miles of land to restore an ice-age ecosystem. In doing so, they hoped to help reverse climate change.

The mammoth steppe was once the dominant Arctic landscape, filled with grazing animals and predators. Grasslands sequester carbon in their root systems. In the winter, herbivores trample snow and keep the ground cold. To that end the Zimovs introduced reindeer, Yakutian horses, plains bison, moose, musk oxen, fat-tailed sheep, Kalmyk cattle, and yaks and are working to introduce carnivores and predators: lynx, tundra wolves, Arctic foxes, brown bears, wolverines, red foxes, sables, and stoats. Allan Savory established the Savory Global Hubs in an effort to restore one billion hectares of land using animals by 2025. These great pasture ecosystems would cool the earth and keep carbon in the ground, where it belongs.

. . .

We cannot feel the future even as it keeps arriving, but we can prepare the ground for regreening. That year of 2007, I felt a great dryness everywhere: the dry meadow bent up into mountains, peak stacked against peak. In the hottest afternoons there was an eerie absence of birds, animals, even mosquitoes and flies. As drought took hold, pond water evaporated, and the days felt weightless, like the recently discovered neutrinos about which a physicist said, "It's as if they want to be nothing and yet aren't allowed to be."

"The effects of climate change are no longer subtle," the climatologist Michael Mann said as global temperatures climbed. In the next days and years, more ice sheets would collapse, and oceans would grow hotter. But late in the day, when it rained unexpectedly, the world around my cabin repopulated. Vesper sparrows sang hidden in sagebrush, a familiar pair of ravens cruised by, tilting and cawing, and the four western bluebirds that hatched out under the lid of my propane tank returned to visit and say a late-fall farewell.

*

Yesterday I was young. Today I'm trying to find a place to live where, as temperature increases, I will be able to find water, grow food, feed animals. I lay out a map of the world and see immediately that the choices are limited, yet I have a choice. Many don't. Soon, perhaps, such a privilege won't matter.

Our deep inflexibility seems a kind of foolishness, or worse, suicide, as we rock between acts of heroism and greed, self-discipline and self-indulgence, and at the drop of a hat we can easily persuade ourselves to go either way.

"I guess we're a failed species," an acquaintance said after exclaiming that there was no such thing as climate change. We're a failed species because despite our unrivaled intelligence, we indulge in delusional behavior to protect ourselves from painful realities. We talk but don't act. We ask to be spoon-fed, but only the things we want to eat, and make demands on the earth without ever inquiring what the needs of the earth might be.

20.

In 2012 I made a brief return trip to Qaanaaq to help the artist Mariele Neudecker photograph there. The ice had come in late and was still three feet thick in May. On the plane north, by chance I sat next to Julian Dowdeswell, the head of glaciology at Cambridge University. He and his team were investigating subglacial lakes and a 460-mile-long canyon that had been recently discovered under the ice sheet. Looking down at rotting masses of ice between Savissavik and Moriusaq, he said quietly, "Ice can always deform. It's too late to change anything. All we can do is deal with the consequences."

In town pairs of eider ducks flew over and thousands of little auks fluttered toward the cliffs at Neqe, where they nested on ledges in rock cliffs. Spirits were high. The ice was still strong, and going to the ice edge where the winter ice had opened up meant there would be fresh meat. I was disappointed to find that Jens had left unexpectedly for a meeting, though he called from Brussels to say hello, but the rest of the family was ready to go.

No one travels light in the spring, and though the ice had come in late, it was still three feet thick. Mikele's extra-long sled carried a skiff plus my friend Mariele; Mamarut tipped his kayak on its side and lashed it to his sled. I rode with him. Gedeon carried his kayak, paddles, guns, tents, and food plus his new girlfriend. The spring snow on top of the ice was wet and the going was slow, but it was wonderful to be on a dogsled again.

I'd dozed off when Mamarut said, *"Hiku hinaa."* We had come to the edge of the ice. Camp was set up. Gedeon sharpened his harpoon and his girlfriend melted chunks of ice over the old Primus stove for tea. The ice edge means life has returned. Pod after pod of beluga whales and narwhal swam by, and the middle-of-the-night sun tinted the sea green, pink, gray, and pale blue.

Gedeon and Mikele carried their kayaks to water's edge and squatted there, waiting. More pods of walrus came by, but the men didn't move: "They had too many young ones with them," Gedeon whispered. When a third pod passed, he climbed into his kayak and exploded into action, paddling hard. Harpoon in hand, he hurled it. Missed. Then he turned, smiling, and paddled back to camp. There was ice and there was time—at least for now—and he would try again later.

In the night a group of Qaanaaq hunters arrived and made camp right behind us on the ice. It's bad practice to usurp another family's hunting site, especially "A" hunters like Jen's and Mamrut's extended families, who were the elite hunters of the northern Arctic. The others should have moved on, but they didn't. No one said anything. Just as the sea ice was disintegrating, so were the old courtesies.

Later, a dog fight broke out at the intruders' camp and an old man started beating his dog with a snow shovel. I stood to run across the ice, hoping to save the dog's life, but Mamarut grabbed my arm. "No. Don't go. It's dangerous. He might hurt you." I sat down and covered my eyes and ears. The sound of the dog yelping was unbearable. Sled dogs are treated with respect—we depend on them to save our lives. In more than twenty years of traveling in northwest Greenland, I'd never seen anyone beat a dog.

Hunting was good the next day and the men were happy to be bringing food back to their families. Though the ice was strong for the time being, they knew better than to count on it. We were deeply upset about the beating we had witnessed, but there was nothing we could do. Uncertainty was taking its toll, and the unwritten codes of honor that, together with the old taboos that had kept Greenland's ice-bound society humming for five thousand years, were eroding. The dog beater would be silently shunned for his violence. The number of dog teams was already disappearing along with the ice, and the villages once filled with all-night howling—what was called a Greenland chorus—were going quiet.

That night Mamarut told us about the village of Moriusaq, where he and Jens had grown up. We had been there in 2004. Now the village was deserted. "There were only two old men left," Mamarut said. "They had lived there all their lives. When they went out hunting together, a gun went off accidentally. It killed one of them. The other man couldn't stand being the last one left, so he killed himself. Moriusaq is no more."

One by one the threads of traditional lifeways were being pulled out as the ice conditions worsened, and signs of cultural collapse had begun to show. "We may be coming into a time when it is all summer," Mamarut repeated as he mended a dog harness with whale gut thread. One of the strongest hunters in Greenland and also the funniest, he was now too bunged up to hunt and too depressed to be funny. He'd broken his ankle going solo across the ice cap in a desperate attempt to find food—he had been hunting musk oxen instead of walrus—and it took two weeks for him to get home and another week before the plane arrived to take him to the hospital in Nuuk for surgery.

His ankle still gave him trouble and his rotator cuff was torn. The previous winter his mother died—she had made all the skin clothing for her four middle-aged sons: polar bear pants, fox fur anoraks, sealskin kamiks and mittens. Recently, a handsome younger brother who lived in another village had committed suicide. Mamarut recalled that the brother "always had everything perfect: his dogs, his skin clothing, how he looked, his hunting. But he was sad when the ice went."

Resilience and *adaptation* are words bandied about by outsiders who know nothing of this life. "They want us to become fishermen now," Mamarut said. "How can we become something we are not?" It was clear to him that *adaptation* was a euphemism for "your culture is fucked."

Climate is culture. The last day we camped at the ice edge, Gedeon and Mikele harpooned two walrus, four narwhals, and ten halibut. As the men paddled back to camp, their sled dogs howled with excitement, knowing they would be well fed, and so would we. But the ice was getting dangerous. Mamarut had stayed in camp to pack. His body was failing as if in sync with the ice, and his brash younger brother, Gedeon, had taken his place.

Once the ice in spring had extended forty miles out into the strait. Now it barely reached beyond the shore-fast ice of Qaanaaq, and despite seasonal fluxes, the ice kept thinning. Mamarut shook his head in dismay: "Ice no good," he blurted out in English, as if it was the best language to express anger and dismay.

On the way home to Qaanaaq the next day, Mamarut fell in the trace lines while hooking up the sled dogs and was dragged a long way before I could get the dogs to stop. I wondered if these might be the final days of subsistence life on the ice and if I would ever have the privilege of traveling with these men again.

21.

I've tried to bore into the heart of things, but some days it's like banging on a metal plate. There is no inside anyway. My feet are restless. They yearn for the ski, the dogsled, the stirrup. Old days I climbed on one horse, then another. Sometimes I rode fifteen hundred miles a year. Winters, I checked heifers on skis and, for twenty-three years, sat with my legs out straight on freight-sized dogsleds, a posture that still feels most comfortable today.

In the 1960s, '70s, and '80s it wasn't Greenland and Africa I traveled to, but Japan. Few spoke English and I could read no signs. Up north I visited *itako*—women who speak to the dead. In Kyoto I spent a month with two women Noh mask carvers, watched rehearsals of Noh plays every afternoon, and attended performances at night. The old houses delighted me, their wide wooden planks peeled back, potted plants set out front, and a tiny altar at the end of each narrow street. Nights, fire guards walked through neighborhoods banging wooden blocks to let residents know they were safe. Afternoons, an old man at the end of the street played the shakuhachi as I strolled in my yukata to the public bath. I loved the paper doors sliding open one after another, an eternity of openings through which one could enter and exit, how all rooms looked the same, the tatami smelling like dried grass, how each room was a field.

Years later I climbed Hokkaido's Daisetsuzan. A volcano is its own desert. The trail was gold, pink, white, purple, and the rocks were black. Islands of vegetation appeared and disappeared. Mountains all around were thick with Sakhalin fir and Ezo spruce. Every step was fire, every other step was the cooling element that hardened the mountain snow that my guide, Michiko Aoki, skied from the highest elevations, leaping over steam vents. A man carrying a white umbrella passed. A red dragonfly flew the other way.

Up top we ate lunch with the other hikers. On the other side, we pitched tents on a patch of dirt that during a midnight rain turned to mud. Wind howled. A thick cloud pushed through as we started out in the morning, down through a clump of fringed gentian and prostrate pines. Dappled sun buckled as mist streamed. We saw the mountain; we didn't see it. The path remained.

Down and down. Fox scat, birch trees, dwarf bamboo. A young man passed and said, "It was so windy up there it blew my mouth open!" Funnier than it sounded because Japanese hikers are silent when they trek. Michiko asked me if it was okay if we turned off the main path and take a little-used trail to a hidden *onsen* where her boyfriend was the bear biologist. "Of course!" I said. "No Caucasians have ever stayed there," she warned me, but once we arrived, the welcome was extravagant. We were ushered into the simple room where the emperor and empress had stayed.

Michiko's boyfriend was tall and ursine. He quickly told us the rules: "From here up, you must stay on the trail. You will arrive at a meadow below the revetment. It belongs to the bears. You must leave in time to be at the bottom by three p.m. That's when the bears come out. They need to be alone; they need their freedom. Mondays, the trail is closed. Mondays are Bear Days: no people at all."

By three in the afternoon, we were down in the parking lot at the inn. A photographer set up his spotting scope and we looked with binoculars. Soon enough, two bears strolled out, as if on schedule, then a sow and a cub. They rambled, lay down, scratched their backs in the grass, looked for berries. When the sun went down, the bears disappeared. Later, alone in the outdoor hot pool, I looked mountainward, wondering what the bears were doing, where they slept.

We seem to know more about how ice deforms and how trees, plants, and fungi communicate than we understand about what animals know. They are too much like us, and we barely know ourselves. When I asked Michiko's boyfriend why the bears behaved as they did, he shrugged. Did they know the bear-human schedule? Did they eagerly wait for us to leave?

"We believe they must have their secrets," he said, "so we don't bother them." I realized then that the purpose of the bear biologist in this hidden place was, charmingly, not to know.

Later, in Sapporo, I asked an Ainu gentleman about bears. The Ainu were the first residents of ancient Hokkaido and revered the Ussuri brown bear. Their knowledge of the ecosystem consisted of dividing the land into "bear fields." He said, "We know the land through the minds of the animals."

22.

Cora. Spring again and the snow on Glover Peak still looked thick. I'd hiked above a remote canyon in the Winds where water churned over broken slabs of granite and swirled into smooth basins. The "churn" was the mana of the mountain. Down-trail, it emptied at my feet, all foam and confusion. The whole world was in there, but when I tried to scoop it up, I lost everything I'd gained. What I couldn't grasp, I dug for avidly, my pickax gouging earth, trying to find the sinuous routes of mind and matter.

A mother pronghorn had given birth to twins just a few yards from my cabin. They are born without a scent so as not to attract predators, but the mother kept a lookout for coyotes, eagles, and wolves anyway. Fifty percent of the young die from predation, but that day I saw only ravens doing scud runs across the meadow.

Up at four thirty a.m., I watched light come as the moon set. Sun and moon were on balancing poles. Earlier the bunchgrass in front of the cabin had set off systemic signals telling plants and waterways everywhere to wake up. First the grass was spangled with frost. Later, it was bound together by floating spider silk. When the pronghorn twins stood, a whole field of long-stemmed blue flax waggled.

Finally, the full sun popped up above the cordillera. The

orb. Bursting. A rolling ball of unfelt heat. Near the cabin soft pine tips, newly grown, lit up like green bulbs.

Driving down the dirt road, I saw something odd: a young pronghorn antelope had its head stuck in a woven wire fence. I stopped; climbed down from my pickup, fencing pliers in hand; and approached slowly, cooing to the frightened animal. The fawn had been there for a while: there were abrasions on her neck and leg. From behind, I pushed my hand through the wire and clasped her chest. I could feel her heart thump. Her tiny brown mane brushed my face.

Gentle, gentle. I didn't want to let go. I looked around: no mother in sight. Perhaps I could take her home, bottle-feed her. I cut the wire from around her head and chest. Cut a large hole while still holding her so she wouldn't lunge ahead and injure herself. She pushed hard against me, wanting to run, and when I let her go, she charged ahead. Just then, a female pronghorn rose out of a swale and stood, alert, watching. The fawn ran to her. When they met, the mother sniffed, jerked her head up to look at me, sniffed again, then licked. Finally, the fawn nursed.

For a long time I watched, unmoving, longing to hold her again. A last look, then the two animals ran. For the rest of the summer, any time I drove the road, I could whistle, and she'd come with her fawn to show me that they had survived.

*

That was the year 330,000 acres burned in the west, erasing grasslands and gnawing into trees' cambium and circulatory systems, right into the phloem and xylem, drinking their blood and bodily sugars. My dog, who I kept thinking was dying, lived on.

Mornings, blackened pine needles covered my outdoor writing desk—dark pins that no longer held things together. A bull moose moaned for a cow moose in heat; pronghorn antelope ran from a hidden wolf, then turned and chased him away. No wildflowers bloomed. In a cold snap, aspens that had leafed out too early froze.

Winters, most of the eight hundred thousand pronghorn antelope went south to the Red Desert, where wind blows the open country free of snow. Sage is the pronghorn's favorite browse and Wyoming has plenty: it contains as much protein as alfalfa.

Spring came with the arrival of mallards, and family groups of pronghorn and mule deer followed, crossing rivers and highways as the snow melted ahead. It's said that pronghorn developed their famous fifty-mile-per-hour speed while running from the now-extinct American cheetah. I watched a group of twenty climb a ridiculously steep hill, then turn around at the top just to run down. The pronghorn's migration trail is the longest in the Lower 48. The whole corridor is a thick strip of land wedged between mountain ranges, from the Red Desert up along the Wind River Range, up and over the Gros Ventres. When their individual routes are plotted on paper, it looks like the strings of an oversize violin.

Shelter in place, we are told when we're in danger. But for the pronghorn and mule deer, "place" is not a dot on the map but a long vista of land rather than a single destination. Cued by vernalization, the animals move with uncanny prescience to wherever the forage has just become prime. They know when and where to find the highest nutrition and arrive on exactly the right day. Biologists call it "surfing the green wave." The whole corridor is a seasonal habitat, and their routes through it change every hour, every day.

The young pronghorn I had saved and watched reunite with its mother would be on the move any day. Our "always in a hurry" ways of living seemed doubly insignificant. Intimacy with weather, terrain, and pronghorn taught me to hold each foot-worn trail in my mind as it deepened. I liked to think that a "green light" glowed inside those animals, instructing them how to survive and eat well. We humans might do the same and, in the process, vernalize our minds.

Snow clouds hid mountains. The pronghorn would have known a week earlier that it was time to turn around and start south. When the sky cleared, Squaretop Mountain appeared to be a huge pin that held me in place. The bed in my cabin was built into an alcove with glass on three sides. From there I could see the stars in the western, northern, and eastern skies. When a meteor flew by, I wondered if the pronghorn had seen it and followed its light; if the meteor was fleeing or had just arrived.

That September I was invited to a party in Cora honoring Neal Conan. He'd interviewed me on his daily NPR show, *Talk of the Nation,* from Washington, D.C., but we had never met in person. I was in a foul mood and went to the party reluctantly. I hadn't bathed. My dog, Gabby, had run off during a thunderstorm and I'd been out looking for her.

When I entered the room, Neal was talking about his years as a war correspondent and casually mentioned that he had been captured in Iraq during the first Gulf War. When the guests wandered away, I stepped forward: "What really happened?" I asked, and sitting knee to knee, we discovered we'd both had close calls with death and dark nights of the soul.

In his essay on friendship, Emerson wrote: "Friendship requires that rare mean betwixt likeness and unlikeness, that piques each with the presence of power and of consent in the other party."

I was piqued, and we had nothing in common. Neal was on the air live for two hours a day, four days a week, but soon enough, he began coming to Wyoming for long weekends. It was a hectic commute: headphones off, he'd race to BWI and catch the last flight to Salt Lake City. At the same time in Wyoming, I'd start my four-hour drive to Utah and meet him just as the plane arrived. With the cold dinner I'd made of artichokes, roast chicken, and cookies for the road, I'd gather him up, turn around, and go north again. We'd arrive at the cabin well past midnight but before dawn.

Sleep didn't matter. Love had given us a youthful exuberance. We'd hike, picnic on the trail, climb a glacial erratic and look at the view. Monday came too soon. I'd drive him back to Salt Lake to catch the last plane east, then turn around and drive home.

We had known each other exactly thirty-five days when he proposed. He'd asked me about the word *forever*. I didn't know what he meant. Then he said, "Would you consider living with me for the rest of your life? Because I already know that's what I want."

I was stunned. Then I shocked myself: I said yes. After, we found ourselves on the floor, rolling, holding each other, and laughing, a deep, roaring laugh of recognition.

Life is absurd; death waits for us; unconditional living is true. So much of our problem is our point of view. The Sanskrit word *prana* defines the point of view of "ground." Not soil but rootedness, knowledge inseparable from reality. The Tibetans call it *sang thal*, "reality's reality simultaneously penetrating."

Three months after Neal and I met, he asked me to come east for the winter to live with him. By the time the plane landed in Baltimore, I felt skittish and apprehensive. The car service that was to meet me hadn't showed. I stood at baggage, waiting for my two large suitcases, quite sure this wasn't going to work out. The romance had gone too fast and he hadn't even bothered to meet me. Immediately I made an escape plan: I'd get a hotel room and book the first flight back to Wyoming in the morning.

There was a tap on my shoulder. In my fury I whirled around: it was Neal, smiling and apologizing as he thrust a bouquet of roses into my arms (though he confessed they were from the airport vending machine), and so, our life together began.

23.

Early September in Wyoming was summer becoming winter. Snow squalls sashayed across the meadow, making me think that all was well with the world, but it wasn't. Yet I had a strange feeling of well-being, of never having been happier. In the southern hemisphere, it was the hot season. The river that went by Allan Savory's camp was dry. Elsewhere the young of kudu, impala, elephant, and giraffe would falter and die. "This is the worst I've ever known in my whole life in Africa," he said. "What we are seeing is extinction in action."

In the far north, the glaciologist Koni Steffen said, "Believe it or not, water vapor is now the most prolific greenhouse gas in the atmosphere." Hot oceans were throwing moisture into the dirty air. Northern regions were getting wetter and snowier and dry parts of the world like sub-Saharan Africa, Australia, and the American Southwest were getting drier and hotter. The North Pole no longer existed in the physical sense. Its disappearance stood for the profound alterations in our lives.

"A small change can have a great effect," Koni said. The frozen seabed clathrates made of mollusk shells that had sequestered carbon for millions of years were beginning to melt. Permafrost was collapsing. Warming and acidifying oceans were causing outlet glaciers to crack and melt at both poles. The Arctic seabed methane pulse and ocean heating were deemed one of the greatest risks facing us. "There will be heating events that devastate agriculture. There will be famine, and many will die," Jason Box said. After two decades of assiduous work and discovery, Jason stepped off the Greenland ice sheet and went to South Greenland to plant trees.

The Arctic Oscillation continued to flip back and forth like a toggle switch: "Live or die," it seemed to be saying. The jet stream meandered so that storms lingered, bringing destruction and death to those below.

No matter what, the morainal humps in my road still threw me: I banged down onto them, skidded sideways, and moved from erratic to erratic trying to see where ten thousand years ago the ice had passed by. Neal and I agreed that I would spend summers in Wyoming while he worked in the east. On nightly phone calls from Washington, D.C., his laughter and brilliance held me as the global whatchamacallit—climate change and its feedback loops—kept writhing. In the summer of 2013, an assignment from a public radio producer took us to Greenland to record the sound of glaciers melting. We sailed north from Ilulissat and arrived at Eqi Glacier just in time to see two panes of ice whooshing straight down and propelling a wave that slammed the side of the boat. A long line of glacial till slid into growing piles. Warming oceans were undermining outlet glaciers. Ice sheets were sliding. Melting sea ice was exhaling methane at the glaciers' margins.

We watched the calving front in silence. It seemed to say everything about climate chaos. The glacier kept calving and retreating, cutting off more and more of its own body as if cutting off limbs. At the end of the day, when the boat turned toward town, a thick fog enveloped us. Inside the cloud it was all glitter and chalk, but when the mist thinned, a fogbow appeared aft. It followed us as if being towed, then vanished, all the promises of a healed world left behind.

There was no doubt that we had to alter radically the habitual practice of human dominance in the natural world. Back home in Wyoming, between my cabin and the meadow that lifts into big mountains, young pronghorn antelope were having running relays on sun-cured grass and sandhill crane chicks had grown tall. Hatches of bugs appeared in swarms, and trout rose to catch them. There is no one keystone species in an ecosystem; all living beings are essential, but too many of our landscapes are half-dead and lack the thrilling liveliness that could be there. Human domination has been ruinous. Plowing turns the world upside down. Overgrazing and undergrazing create deserts. Chemical fertilizers make the farm field an addict.

If the two root causes of climate change are loss of albedo and land degradation, we have to find a practical way to cool an overheating world. Wild grasslands sequester airborne carbon in their roots. To stop climate change we have to reinvigorate as much of the damaged planet as we can, the natural way. Nicole Masters, a soil ecologist from New Zealand, said, "We can't reverse the extinctions that have happened, but we can reverse abundance decline." A billion hectares of land under Holistic Management by 2025, ten thousand reindeer on the sacred land of Sapmi, and a greatly expanded Pleistocene Park, with its massive grasslands and mobs of grazing animals carrying the ancient bloodlines of bison, horse, cow, deer, and elk that still know how to survive. If the square miles of grasslands stack up, a panoramic liveliness will break out and climate chaos will be averted. Every blade of grass counts if we are to survive.

. . .

Half a mile from my cabin I sat on a huge boulder—a glacial erratic left behind by retreating ice—and looked out over the range, then held my head in my hands. A neurologist explained that the balance between stability and mental chaos occurs when a phalanx of neurons fires, triggering others, until across the whole brain they cascade and avalanche in sequenced, tidal rhythms.

Quiescence and fever. Starvation and extinction. Joy and blight. Ecosystems kept collapsing in sight, out of sight, and I had to work hard to remember that loss and abundance co-exist, and both are true.

24.

Cooper Island, Alaska, is a radical place—radical in that it hardly exists and could be wiped away by a few rogue waves. A remote, windswept barrier island twenty-five miles east-southeast from Utqiagvik (formerly Barrow), it has been the summertime perch for ornithologist George Divoky since the early 1980s. His forty-five-year-long study of the ice-evolved Mandt's black guillemot has become an unrivaled index of abrupt climate change.

The August day Lonny, a local from Utqiagvik, took me by boat to the island, the clouds were gray and flat, but the wind was rising. Once aboard, he threw a life jacket to me, then revved the two big engines. We roared across Elson Lagoon. The trip took two hours. On approach the profile of the island was so slim, it was difficult to see. Lonny pointed at a brown splinter: "There it is. Home sweet home."

George was at the shore waiting. Mop-haired and grizzled, his handsome, lined face collapsed into a wide smile as Lonny cut the engines and we bumped onto the sand. "Welcome to Cooper Island," George said. We were old friends and, arms out, he hugged me against his black rain bibs splattered with liquid bird shit and held up by orange suspenders.

The island is a flat, three-mile-long slab of cobble, gravel, and sand with an elevation ranging between two and six feet.

On one side is Elson Lagoon; on the other is the heaving Beaufort Sea that has torn Cooper's lithic boundary into misshapen fingers and pocked its torso with saltwater ponds where storm-driven ocean waves have intruded.

On that first blustery day George's barrage of conversation began, his overlapping thoughts igniting and dying out like sparks. He lugged my two bags—one filled with camp food, the smaller one with a change of clothes—toward a plywood shack half a mile away.

The gravel was strewn with groups of black Pelican cases, the waterproof, shatter-proof cases used by photographers around the world. "Those are the nests," George said. They're polar-bear-proof. He'd drilled three-inch holes on the sides of the cases—big enough for the chicks to go in and out, but too small for a polar bear to reach in with his paw.

Several times George stopped and lifted his binoculars to glass far reaches of the island. He wasn't looking for birds. Since the retreat of the pack ice in 2003, polar bears had begun frequenting Cooper Island. That first year without ice, fifty bears appeared in the town of Barrow. "One came here and ripped open a tent. I tried to drive him to the north side of the island, but he came back. As he came for me, he tripped on the guy wire of the weather port and was so surprised that thereafter he left me alone."

We kept walking. The Beaufort Sea was whitecapped, its waves crashing against a wall of gravel. "It used to be all ice out there," George said. "Very still and quiet, and there was ice right to the island, even at the end of summer. Now the sea ice is two hundred miles away."

We stepped over lines of driftwood that were strewn across the island, plus the detritus from a cold war camp made by the Navy in the 1950s. Pairs of guillemots, black bodied with white wings and bright-red feet and legs, fluttered and waddled almost comically. "They are the only ice-evolved seabird in the far north, other than ivory gulls," George said. Birds flew low and fast a few feet from us and returned with small fish clasped sideways in their beaks. These they delivered to the hole on the nest boxes for their chicks to retrieve.

Guillemots are part of the auk family. They can dive, swim underwater, and hold their breath for almost three minutes, but are constrained by water temperature from successfully catching fish. In warm water, fish swim too fast for the birds; they need the slower Arctic cod at the ice edge in order to feed themselves and their young. "Now the cod is gone because the ice is gone," George said. "And the guillemots have been reduced to fishing for sculpin in the lagoon."

We walked by a pair of birds that had lost their young: they stood idly by the nest boxes as if waiting for a miracle. Males fought, jumping up and down and squawking; others lined up at the pond—their swimming hole—and joined red phalaropes, dunlins, and long-tailed ducks.

"Every year for forty-some-odd years I've come to Cooper in June and try to time it so I can watch the guillemots arrive. If they come too soon and there is snow on the ground, the predators will get them—black bodies on a white ground. The season is short. As soon as the snow melts, there's the whole performance of copulation. The males stand up straight with beaks tucked in and strut around with big steps. Everything gets quiet during incubation—males and females take turns. After the chicks hatch, there's lots of flying out to sea to get food. Two months after hatching, the chicks are ready to fledge. It's a quick turnaround."

Clouds sped overhead as if mimicking George's rapid-fire mind. "I came here by accident," he told me. "I was doing surveys for the Smithsonian in the late 1960s and saw ten pair of guillemots here. There were lots of predators and no truly safe place to nest, but when I came back a few years later, they were still here, so I decided to stay. I mean, what's the point of being an ornithologist and having a government job or getting stuck in academia? All those meetings. They miss the life of the things they are studying. I wanted to live with the birds."

We slipped through an opening in a three-strand electric fence turned on only at night. The "compound" consisted of a tiny weathered plywood cabin, a large tent, a storage tent, a wind turbine and a weather port, and a few solar panels.

"The electric fence doesn't keep the bears out," George said, grinning. "It's just a reminder to them that they share the island with me. You can throw your bedroll into the tent. That's where you'll sleep. And the bathroom . . ." He turned and pointed. "I call it 'South Beach.' Just go to water's edge behind the cabin. No one but the birds will see you."

We stepped across the pallets stacked with unwashed pots and pans and up two stairs into his summer abode. It was cluttered and disheveled with the strong scent of the unwashed. "We have to haul water here now that the ice is gone. No sense wasting it on bathing." George peeled off his brown jacket but kept the rest of the layers on—a Nano Puff jacket and a sweatshirt hoodie. The yellow field books where he records his data were neatly stacked on a shelf above his platform bed, and a row of hooks held headlamps, binoculars, cameras, and hats. Three guns leaned against shelves stuffed with boxes of ammunition and camp food: peanut butter, instant soup, ramen noodles, cans of dehydrated meat and vegetables, soy sauce, hot sauce, and chocolate.

He lit the Coleman stove, opened a sliding window, and put water on to boil. "Don't get the wrong idea," he said, laughing. "I'm not some kind of Arctic explorer. I was a kid from suburban Cleveland—I'd never even been camping."

We drank tea from barely rinsed coffee cups. "This island

has been ideal for the long-term study of a seabird because it's remote with a minimum of variables, and nest fidelity is very strong with guillemots. If they survive, they come back. That's what makes the data meaningful. In June I know immediately which ones survived the winter, which ones died. I have the exact counts and continuous data sets. No one else does. Most studies last only five or ten years. But they would have missed the whole arc of the changing climate. I have incontrovertible evidence."

He poured more coffee and unwrapped a large block of chocolate. "It's not just about the birds," he began. "It's also the island. When I first came here the island was surrounded by multiyear ice and seasonal sea ice. You could always see it. The guillemots fed in open leads in the ice, eating Arctic cod. Everything was eating it: bears were eating seals that were eating cod, and the birds fed it to their chicks, and we ate it too. Everything was a repackaging of Arctic cod. All of us were sea-ice dependent."

After tea we went out to do nest checks. At a "colony" of nest boxes grouped together on the west side, George carefully opened a lid and gently held a chick. Placing the bird inside a small nylon sack, he weighed it on a hanging scale, measured one wing, and put it back through the opening on the side of the plastic box. Then he entered the data in his yellow notebook. On it went from box to box, twenty-five in all.

"I know these birds. I know who this chick's parents are," he said, holding another bird up to be weighed. "Every bird since 1972 has been banded—that's seven generations of birds. There was a chick born in 1991. She was gone for a long time but came back. I'd love to know where she'd been. Now she's back and she's breeding."

Mist surrounded us, but as we walked into it, it dissipated as if giving us more room on the ever-changing boundaries of the island. A guillemot flashed by carrying a tiny sculpin sideways in its beak. "It's an ugly fish, probably not very tasty," George said. The parent lay the catch beside the nest box, then flew away for more.

At another box the chick George weighed had lost thirty-five grams. "Don't think she's going to make it," he said, returning the chick to the nest. The weather turned blustery. It was thirty-nine degrees but felt like twenty. George opened another box. "There used to be two chicks in every nest. Now only one. The aggressive one gets all the food. I note down the time it takes for the sibling to die. It's pretty depressing. Brood reduction: never saw it before 2003. I look into the eyes of a dying chick that I know will be dead the next day. That makes my experience of climate change different from most people's."

He stood up and turned into the wind. "I've watched the multiyear ice on the Beaufort disappear. I've watched the guillemot population crash. These are real deaths, not just something you read about, and it has affected me."

He pulled up his hood and, rocked by wind gusts, walked to the easternmost colony. "The oceans are getting hot. Deep water was found to be twelve degrees hotter than it was nine years ago. There are toxic hot spots in the Beaufort Sea, and the weather is different: we have clouds and rain we never had before. People are surprised I keep coming back. I'm surprised that people don't return to the birds they've been studying."

More chicks were weighed and measured, and the data logged in. He closed the yellow notebook and held it solemnly. "The truth is, my study of the guillemots has become a study of how extinction happens. In 1989 there were 225 breeding pairs here. We started this year with 75. Now there are only 35 adult guillemots and each pair is only able to raise one chick—and of those, there are only 25 chicks left. When I get up in the morning I don't know if they've gained or lost weight, if they're going to die or fledge. And frankly, I'm not sure if I want to spend the next ten years watching the population get smaller and smaller. But if I wasn't here, I'd be constantly wondering what was going on. You miss one year and the whole data set loses credibility."

Nearby another pair of guillemots stood by a nest box where they'd lost their last chick. They kept looking in to see if, miraculously, there was a baby. "It's way too late to be raising a youngster, but they're hoping, grieving, wondering, much like all of us are."

Ever vigilant, George glassed the island. "A few days ago, there was a polar bear sleeping on the grassy place between the cabin and the shore. Pretty close. Now he's gone. Maybe I've been out here alone too long, but I feel bad about him leaving! Didn't he like me?" He laughed, then pointed at a jaeger swooping across the midsection of the island. "That bird waits for someone to bring a fish in, then steals it!"

That afternoon we saw a white lump at the far eastern end of the island. George set up his spotting scope and we took turns watching the polar bear sleeping, sitting up and looking around, then lying down again. A few moments later, we spotted two more bears: a mother with a yearling cub that crawled across her belly to nurse, then fell back against her and slept.

"I used to store my food in an ice cellar," George said. "Everyone in Barrow did. But no more. The top part of the permafrost has melted. A long time ago there was a winter camp out here. Archaeologists found houses made of driftwood and whale ribs with a whale skull used as a step down into the excavated interior. Heat and light came from seal-oil lamps. The house was square but banked with sod, and if the sod softened, they poured water on it until it froze. Now we store our food in beer coolers."

In the cabin, George pushed back on his platform bed and leaned against the wall. He's inured to discomfort and cold. "Last winter, a polar bear tried to break into the cabin. He made a hole with his claw through the wall but couldn't get his head in. Too big! So he gave up. When I'm here my smell keeps the bears away," he said, smiling.

Cold night. We went out to do a last nest check. The wind howled, and cresting waves looked like galloping polar bears. A gust knocked George forward. He shrugged, laughing. "It's not as bad as it looks," he said. "You know how it is—this addiction to the high Arctic—you have it too! We call it 'the Arctic high.' Pure serotonin. See, I don't want to go to town for a hot meal and a shower. Just walking by a nest box and knowing the whole guillemot family, that's enough for me. I can't explain how completely rejuvenated I am by my three months here. It keeps me going for the rest of the year."

All night the Arctic Ocean churned like a wheel stuck in place. With no lid of ice, the open water was ferocious. Retreating ice reminded me of a horse that can't be caught, moving away every time one comes close. The Salish word *sumic* means "sacred flow, the intertwining of nature and lifeways." How much of that was left? Twice a day every day for three months, George kneeled, weighed, and measured chicks with gestures so redundant they began to look ceremonial.

At seven a.m. guillemots waddled and fluttered, their red legs bright against pale sand. The Beaufort had pushed mist upward into a barely perceptible dawn. Walking east, we spotted what George called "the ghost tracks" of polar bears made during the night. They shone white in the morning light.

Nearby a family of brown-feathered brant ate grass so slender it was almost invisible. How many green threads did it take to keep them alive? Soon the brant would fly south from Cooper to the Izembek Lagoon in the Aleutian Islands, fill up on eelgrass, then go south along the Pacific coast to Mexico.

George looked at them smiling. "They're part of my family out here: brant, guillemots, and polar bears." As we walked back, purring guillemot whistles accompanied us. He scanned the edges of the island. "Even if the guillemots adapt to an ice-free environment and raise only one chick, the rising sea level will someday take this island. The rules out here are strictly Darwinian. That applies for me too."

Back inside we warmed ourselves with a dram of rum as George pondered moral questions: "Should I start supplemental feeding to increase fledging success? Because I've already interfered with the natural order of things by bringing plastic nest boxes to the island. Where do I stop?" He poured more rum into my water glass. "For two decades, Cooper was a good place to breed. Now there are fewer and fewer fish to feed the young. I've provided housing for the birds, but the project can't be sustained. It's a bit like Western civilization: it looks good, but it's dying."

I stepped outside, needing to breathe. It was mid-August and the air was cold. Fifteen minutes of daylight was being lost every day. "I worry about the Arctic being opened up to drilling," George said, joining me outside. "Capitalism at the expense of nature. Deadhorse, Prudhoe Bay, and ANWR. They want to drill nearby; they want to drill everywhere."

We went to bed early. My tent was large enough to stand in, but it had no windows. During the night I heard noises. George had given me a small contraption with a button that, if pressed, would made a noise terrible enough to scare away a bear. "Well, that's the theory," he'd said, grinning.

Fully clothed—not because I was afraid but because I was cold—I zipped my sleeping bag to my chin. More noises. I lay still and tried to figure out what was going on. We'd left food in a duffel bag outside and unwashed dishes. Had we been too nonchalant? It was probably a polar bear, but what the hell. Better to stay put and relax. I quickly fell asleep.

In the morning, George stuck his head out and whispered to me to come quickly. "Hurry. Come see!" Seventy-five yards away from my tent was another white lump: a nine-foot-long polar bear lying on his side, head resting on his forearm, sleeping peacefully.

I couldn't help but smile. In the presence of bears I feel oddly calm. "It's a slumber party," I said. "Now we have three bears," he replied with a sigh of contentment. "Being alone here I see things in a different light. It's been a strange summer. The sea ice didn't retreat; it disappeared overnight. Vanished. The weather changed. More moisture in the air meant that I rarely saw the sun. There's been unprecedented rain here in what's supposed to be a polar desert. But strangely I'm happier here now than ever. I flog myself pretty hard, then I stop and walk around. You have to care about a place. You have to keep going back every day, every week, every year."

In the weeks to come twenty-four chicks would fledge, with seven more on the verge. Black guillemots fly off at the darkest time of night and quickly move offshore to reduce the risk of predators. "But this year, even though hatching success was good, only twenty-five percent of the one hundred thirty-five nestlings survived." George said. "Most of them died in late July and early August when the ocean water warmed so suddenly, even the nearby sculpin were scarce. It wasn't polar bear predation—the chicks starved to death."

We strolled to the far west end of the island for another round of nest checks. Many boxes were empty. A small pond winked in sun, then went dark. Though George was endlessly chatty and voluble, he said he prized his months alone. "I'm happy here, but frankly, I'm also struggling: the work I do on Cooper Island seems increasingly important because of climate change, but in the larger picture, increasingly unimportant."

One evening he walked across the island's narrow waist. Stones rattled as waves broke. Sun hit a log washed up from some distant place where trees could grow. Beyond were the exhausted polar bears that had to swim from the ice for hundreds of miles just to rest on this island.

George kneeled, opened a nest box, and held a frail chick in his hand. "This is one that didn't gain weight. I don't think he's going to make it." He touched the downy head of the dying bird and looked past the sleeping polar bear, beyond the island's fringe. "The view from here," he said, "is a view of the end of the world."

25.

"The past is all deception, / The future futureless," T. S. Eliot wrote in "The Dry Salvages." Memory is a kind of conglomerate of fictive fragments and bits of truth; and the future, being nonexistent, is only a projection with no ground.

Life's wounds don't fester but are moved peristaltically: a pilgrim's progress of sorts as we rumble along. The impact of climate chaos kept drilling into me as I witnessed cultural death spirals and animal extinctions. On my last visit to Copenhagen, the glaciologist Jason Box said, "Fuck it. It's worse than I thought. The trouble with the scientists who make the models and predictions is that they don't spend any time on the ice. You have to have the feeling of the ice inside you to know what's going to happen."

Allan Savory kept trying to dig me out of my despair: "There's no need to be a victim of circumstance or the current doomsday scenario," he said, and suggested I visit some of the Global Savory Hubs whose joint mission was to put a billion hectares into holistic management by 2025.

In northern Sweden I met Jörgen Andersson, head of the Nordic Hub. He's a tall, soft-spoken farmer-philosopher, a twenty-first-century pastoralist who calls his regenerative ag project Mountain Graze. He owns no land, not even the simple house in which he lives, and moves his thousand sheep across unused pastures in the village and in the mountains. He sells meat and wool to townspeople but insists they contribute money at the same time. "That way they become co-owners of the land and the animals they eat. The sheep become ambassadors of a new way of cooperating."

His many projects include a portable milking parlor for dairy cows, portable chicken coops, and vegetable gardens. He designed and built two unique tent-like winter shelters for his sheep. With the Norwegian Hub teacher, Anders, and his Sami neighbor, Per Erik, he is attempting a grand collaboration with a hydroelectric company to put ten thousand reindeer on its unused land. "We are thinking 'upstream,'" he and Anders told me. "With reindeer grazing above the dams, we could wake up the water cycle, stimulate the growth of grass, improve their water storage capability, and help our Sami neighbors recover their traditional herding life, plus draw down airborne carbon."

When the son of one of Jörgen's neighbors came to visit and insisted on lunch, Jörgen made pancakes. The recipe: "gobs of butter, pieces of cooked bacon, a lot of eggs and milk, and a little bit of flour." Flipping pancakes effortlessly, Jörgen said: "Gretel, I believe in our species. I prefer to be positive. You should too. We can make choices and decide what kind of future we will have. Our ecosystems are underperforming. We are standing in the way of nature. I am working to enable the highest imaginable vitality of every ecosystem and bring the deadened earth back to life."

*

Home again. It felt good to stay in one place for a while. I listened with my stethoscope to the urgent respiration of the soil. It seemed I had ears in my feet as I traipsed across sentient ground, hay stubble and sagebrush, and tried to feel the subtle shifts of watercourse, fungi, and filament that connect everything under us.

"As soon as our feet touch the soil, the plants know we are there," a neighboring rancher said. "They know it every time. As soon as I stopped abusing the soil, the land began to wake up from a long sleep, and me with it. If we can put enough carbon back into the earth, the climate nightmare just might slow down!"

I lay on my cabin floor as a fast-moving storm pressed thin layers of wind-clouds together like the pages of a book. Then one snip of a rainbow pushed up from behind a tree. I drove to the trailhead and walked into the mountains. *Sentience* and *sunderance:* those were the two words on my mind. I climbed past Douglas firs, through willow thickets until the revetment appeared, its rock face black from meltwater. Around me exposed tree roots—mountain bones—reminded me of early dogsleds in Greenland made of whale bones, antlers, and narwhal tusks lashed together with sealskin thongs. How noisy they must have been!

Here, the sound of water dripping was the noise of time. I picked the last wild rosehips to quench my thirst. We are an endangered species and live under a dying sun, yet we keep gulping down the nectar of natural beauty as if it had no end.

I climbed over tumbled boulders, around a landslide, and followed a stream. Against an uneven wall of granite, one thin waterfall joined another like a vine. Up I went, grabbing handholds. To climb against the flow invigorated me: time unraveled, and the rush of water and my upward push kept me going. Existence is torque; friction is energy. I edged into the falls, fingering slime, and skittered alongside it until I stepped up onto a flat stone.

There, before me, was a tarn surrounded by pines. I saw

where water spilled over a flaring rim like the bowl I'd bought in Japan: thin walled, rough textured, with a bent lip from which liquid could be poured. The pond reflected two high peaks. At night it might have held the moon. In its waters time pooled quietly before leaving again. The pond saw all, accepted all, gave all. It did away with false narratives. Pond was death, pond was life; death and pond were space. Overhead, a rib cage of clouds passed, a new body assembling itself and dissolving simultaneously, growing old, being born.

I reached the cabin late, stuffed paper and kindling into the woodstove, lit a match, and waited for the fire to warm me, then took out the box where I kept my beaded Shoshone dance moccasins. Made of smoked elk hide, they still smelled good. In a stack of CDs I found Kevin Volans's *Hunting: Gathering*. The music is spacious, sunlit, molten, alive. I slipped the moccasins on.

Two hundred years ago, when smallpox devastated whole Native American communities, a Mandan elder said, "When we know we are going to die, we dance." I pushed off from the log wall and spun across the room. The choreographer Pina Bausch said, "Dance, dance, otherwise we are lost." Outside, a ground blizzard swept through. Snow riprapped frigid air, erasing what I knew and all I feared. *Shelter in place*. I kept coming back to that phrase. In a tsunami, it meant death; in a pandemic, you might survive. But I'd seen how "place" can move on the path of the pronghorn, on the path of ice, and I wanted to move with it.

I thought of the red bridge in Japan's Dewa Sanzan mountains that was said to connect heaven and earth. How, almost forty years ago, I walked over it, turned, crossed again, then headed into a climax forest of cedar trees where the poet Matsuo Bashō and his companion Sora had lain. In the temple complex up top, the wood was sanded down to a smoothness as if discarding undisciplined needs, and rooms stripped of ornament showed the path from which it is possible to see how the ancient, inexhaustible fecundity of any place—whether at the Haguro-san Shrine or out on the range—can be discovered anew again and again.

From the Tale of Genji scroll fragments I saw in Kyoto, I learned *fukinuki yatai:* how, if the roof is blown off, one can see down into people's lives without entering the room; how the "sacred flow" and the "everywhere and all that is" lies shimmering inside us and all around.

I pull the rope of my door tighter
and stuff the window with roots and ferns.
My spirit is turned to springtime.
At the end of the year there is autumn in my heart.
Thus, imitating cosmic changes,
my cabin becomes the cosmos.

—Lu Yun

AFTERWORD

The writing of *Unsolaced* began in the spring of 2017 as a bookend to *The Solace of Open Spaces*, which was published in 1984. The fuel for writing *Solace* had come from the loss of a loved one and the discovery that my heart's home would always be Wyoming, a home on the range for a wanderer. Little did I realize that as I finished this new book, another similar kind of loss was at hand.

Unsolaced began with two words, *sentience* and *sunderance:* how we know what we know, who teaches us, how easy it is to lose it all. I could not have imagined that in thirty-six years the climate crisis would tip us toward extinction, that for a time, a heartless tyrant would rule our nation, that a global pandemic would threaten our lives.

I traveled widely, had boyfriends, sold houses, built houses, cowboyed for friends, and when my horses and dogs died of old age, I replaced them. I could not have imagined falling in love with a man who didn't know how to saddle a horse or harness a team, but I did. Neal Conan's encyclopedic memory and grasp of geopolitics were astonishing; his sweet generosity, sense of respect, and unfailing humor lured me in.

After ten years together, with as much time as possible spent in Wyoming, Montana, and Alaska, we married in a brief lunchtime ceremony at our house in Hawaii, then boarded the red-eye for Rochester, Minnesota, where he was scheduled to have brain surgery at Mayo Clinic. One surgery turned into three, a Christmas Eve medevac flight organized by his neurosurgeons saved him, and proton beam radiation gave him more time to live.

Ranch life had prepared me for acting on the spot, but I had not expected the ferocity and devotion I would experience while taking care of Neal. His four teams of doctors and nurses at Mayo Clinic worked cooperatively, not unlike the Inuit villagers and Wyoming ranchers I've known. I saw how cooperation erases the dividing lines between nature and culture, human thought and natural fact, dying and living.

Since *Solace* was published thirty-six years ago, everything and nothing has changed. Ecosystems are crashing. Terrorism sprouts and vanishes with devastating effect. Coronavirus is on a rampage, reminding us that the roulette wheel still spins. As the pandemic spreads, animals wander through empty cities as if to say that we humans have been in the way all this time. Finally, the sharp lessons of impermanence I learned while writing *Solace* still hold true: that loss constitutes an odd kind of fullness, and despair empties out into an unquenchable appetite for life.

What I have written is an odd kind of memoir, notable—if at all—for what has been left out.

Acknowledgments

Deep thanks to all those special friends who, one way or another, saw me through the years during which I wrote this book.

In Wyoming: Mart, Mike, and Timmy Hinckley; Maggie Miller; Courtney and Maria Skinner; TJ; Jim Nelson; Stan and Mary Flitner; and J. J. Healy. In Montana: Roger and Betsy Indreland, Anne Indreland, Kate Indreland, Deborah Butterfield and John Buck, Tom and Laurie McGuane, Tom and Patty Agnew, Nan Newton and Dave Grusin, Bruce Townsend, Mallory and Diana Walker, Nicole Masters, Jim Yost, Carly Fraysier, and Holly Hatch. In Greenland: Jens and Ilaitsuk Danielsen, Mamarut and Tecummeq Kristiansen, Gedeon Kristiansen, Mikele Kristiansen, Hans and Birthe Jensen, Ikuo Oshima, Dennis Schmidt, Mariele Neudecker, and Aleqa Hammond. In Africa: Allan Savory, Jody Butterfield, Elias Ncube, and Huggins Matanga. In Kosovo: Dr. Rifat Latifi. In California: Nita Vail, Jesús, Pancho, and Bill. In Alaska: George Divoky, Catherine Smith, Greg Dudgeon, Dick and Gretchen Peterson. In Hawaii: Jileen and Richard Russell, Zac Russell, Eddy Pollack, Dr. Melissa Gosland, and Dr. Ravi Kumar. Special thanks to Richard and Kristina Ford, JJ Sutherland, Arline Sutherland, Clemma Dawsen, Fae Myenne Ng, Charlie Ramsburg, Steve Gwon, Galen Wood, David Jacobs, Michiko Aoki, Joe Riis, Jörgen

Andersson, and Nic and Taylor Sheridan. Many thanks to my editor, Dan Frank, to all those at Pantheon, including Vanessa Haughton and Louise Collazo, and my agent, Liz Darhansoff. The deepest kind of thanks possible goes to my husband, Neal Conan, always and forever.

A Note About the Author

Gretel Ehrlich is the author of *This Cold Heaven, The Future of Ice, Heart Mountain, Facing the Wave,* and *The Solace of Open Spaces,* among other works of nonfiction, fiction, and poetry. She is the recipient of a Guggenheim Fellowship, a Whiting Award, the Henry David Thoreau Prize for Nature Writing, and the Harold D. Vursell Memorial Award from the American Academy of Arts and Letters, among other awards. Ehrlich studied at Bennington College and UCLA film school. She divides her time between Wyoming, Montana, and Hawaii.

A Note on the Type

This book was set in Scala, a typeface designed by the Dutch designer Martin Majoor (b. 1960) in 1988 and released by the FontFont foundry in 1990. While designed as a fully modern family of fonts containing both a serif and a sans serif alphabet, Scala retains many refinements normally associated with traditional fonts.

Composed by North Market Street Graphics,
Lancaster, Pennsylvania

Printed and bound by Berryville Graphics,
Berryville, Virginia

Designed by Soonyoung Kwon